The Micro-Doppler Effect in Radar

Second Edition

DISCLAIMER OF WARRANTY

The technical descriptions, procedures, and computer programs in this book have been developed with the greatest of care and they have been useful to the author in a broad range of applications; however, they are provided as is, without warranty of any kind. Artech House, Inc. and the author and editors of the book titled *The Micro-Doppler Effect in Radar, Second Edition* make no warranties, expressed or implied, that the equations, programs, and procedures in this book or its associated software are free of error, or are consistent with any particular standard of merchantability, or will meet your requirements for any particular application. They should not be relied upon for solving a problem whose incorrect solution could result in injury to a person or loss of property. Any use of the programs or procedures in such a manner is at the user's own risk. The editors, author, and publisher disclaim all liability for direct, incidental, or consequent damages resulting from use of the programs or procedures in this book or the associated software.

For a complete listing of titles in the
Artech House Radar Library,
turn to the back of this book.

The Micro-Doppler Effect in Radar

Second Edition

Victor C. Chen

ARTECH
HOUSE

BOSTON | LONDON
artechhouse.com

Library of Congress Cataloging-in-Publication Data
A catalog record for this book is available from the U.S. Library of Congress

British Library Cataloguing in Publication Data
A catalog record for this book is available from the British Library.

ISBN-13: 978-1-63081-546-2

Cover design by John Gomes

For accompanying software, please go to:
http://us.artechhouse.com/Assets/downloads/chen_546.zip

© **2019 Artech House**
685 Canton St.
Norwood, MA

10 9 8 7 6 5 4 3 2 1

Contents

v

Preface

My early study on the micro-Doppler effect in radar was done under an Office of Naval Research (ONR) research project managed by William J. Miceli. During a technical interchange meeting held at the U.S. Naval Research Laboratory (NRL) on November 9, 1995, one of the subjects at the interchange was an investigation on radar micro-Doppler features. I was fortunately invited to attend the technical interchange meeting and assigned to examine radar data containing micro-Doppler features, to investigate the mechanism responsible for the features, and to explore methods for incorporating recognition of micro-Doppler features.

Two years later when I was working on applications of joint time-frequency analysis to radar signal and imaging, I received a set of radar data collected by the Norden Systems (Westinghouse) using a radar prototype. The target was a human being walking toward the radar. From a sequence of range-Doppler images of the walking human, I could clearly see a hot spot of the human body as well as Doppler smearing lines around the hot spot caused by the articulated micromotions of arms and legs. To exploit the details of the swinging arms and legs, I applied a joint time-frequency analysis to the radar range profiles around the smeared portion. As an immediate result, the joint time-frequency representation clearly showed time-varying Doppler oscillations of the arms and legs around the Doppler shift of the human body's translational motion. That was the first micro-Doppler signature of a walking human that I analyzed and represented in the joint time-frequency domain. The signature clearly shows the time-varying Doppler distribution caused by micromotions of such body parts as feet, hands, arms, and legs.

After a decade's work on the investigation of micro-Doppler features in radar, in 2010 I decided to write a book to offer my theoretical and analytical experience on the micro-Doppler effect in radar. The primary purpose of the book was to introduce the principle of the micro-Doppler effect in radar and to provide a simple and easy tool for simulation of micro-Doppler signatures of radar signals reflected by objects with micro motions. The first edition of the book was published in 2011, along with an accompanying DVD to provide MATLAB© source codes, for readers who were interested in simulations of micro-Doppler features. Based on the basic principle and simulation examples provided in the book, readers could make modifications and extend to other possible applications.

Recent developments in the micro-Doppler effect in radar have made more emerging applications and new advances feasible and available. In this second edition of the book, I have corrected typos and errors in the first edition, updated some recent research on radar micro-Doppler signatures (such as quadrotor UAVs, reentry vehicles, and some human activities), and added three new chapters on vital sign detection, hand gesture recognition, and micro-Doppler radar systems. The last two chapters in the second edition are updated Chapters 5 and 6 of the first edition of the book.

For educational purposes, this second edition of the book provides more and updated MATLAB source codes online for downloading. Part of the MATLAB source codes are attributed to the efforts of my students. The source codes are provided by contributors on an as-is basis and no warranties are claimed. The contributors of the source codes will not be held liable for any damage caused.

At the publication of the second edition of the book, I wish to express my thanks to William J. Miceli, my longtime friend and the sponsor of my micro-Doppler research, for his constant support and helpful discussions for the research work on micro-Doppler features in radar, especially technical discussions on basic concepts of the micro-Doppler effect, on angle-cyclogram patterns, and on aural method-based discrimination.

I also give my thanks to Raghu Raj for his work on the physical component-based decomposition method and related figures. I especially thank David Tahmoush for his interesting work on micro-Doppler signatures of quadrupedal animal motions and related figures.

I would like to express my sincere thanks to my students Yang Hai and Yinan Yang for contributions to MATLAB source codes.

Since the first edition of the book, I have received questions and suggestions as well as corrections on the text and on the MATLAB codes. I truly express my thanks to those who made corrections and suggestions, which helped me with the second edition.

I am grateful to the reviewer of the book for comments and constructive suggestions, and to the staff of Artech House for their interest and support in the publication of the book.

The accompanying MATLAB software is at: http://us.artechhouse.com/Assets /downloads/chen_546.zip. There you will also find a list of MATLAB source codes/data that will explain where to download these sources codes.

1

Introduction

In a monostatic radar, in which the transmitter and receiver are collocated, the radar transmits an electromagnetic (EM) signal to an object and receives a reflected signal from the object. Based on the time delay of the received signal, radar can measure the range of the object. If the object is moving, the frequency of the received signal will be shifted from the frequency of the transmitted signal, known as the Doppler effect [1, 2]. The Doppler frequency shift is determined by the radial velocity of the moving object, that is, the velocity component in the direction of the radar line of sight (LOS). Based on the Doppler frequency shift of the received signal, radar can measure the radial velocity of the moving object. If the object or any structural component of the object has an oscillatory motion in addition to the bulk motion of the object, the oscillation will induce an additional frequency modulation on the reflected signal and generates sidebands about the conventional Doppler shifted frequency caused by the translational motion of the object. The additional Doppler modulation is called the micro-Doppler effect [3–5].

Doppler frequency shift is usually measured in the frequency domain by taking the Fourier transform of the received signal. In the Fourier spectrum, the peak component indicates the Doppler frequency shift induced by the radial velocity of the object. The width of Doppler frequency shifts gives an estimate of the velocity dispersion due to the micro-Doppler effect. To accurately track the phase variation of the radar received signal, the radar transmitter must be driven by a highly stable frequency source to maintain a full phase coherency.

The micro-Doppler effect can be used to determine kinematic properties of an object. For example, the vibration generated by a vehicle engine can be detected from the surface vibration of the vehicle body. By measuring the micro-Doppler characteristics of the surface vibration, the speed of the engine can be measured and used to identify a specific type of vehicle, such as a tank with a gas turbine engine or a bus with a diesel engine. The micro-Doppler effect observed in the radar received signal from an object can be characterized as its signature. Thus the micro-Doppler signature is the distinctive characteristics of the object that represents the intricate frequency modulation generated from the structural components of the object and represented in the joint time and Doppler frequency domain.

1.1 Doppler Effect

In 1842, the Austrian mathematician and physicist Christian Johann Doppler (1803–1853) observed a phenomenon on the colored light effect of stars [1]. The apparent color of the light source is changed by its motion. For a light source moving toward an observer, the color of the light would appear bluer; while moving away from an observer, the light would appear redder. For the first time, the phenomenon, known as the Doppler effect, was discovered. The effect states that the observed frequency (or wavelength) of a light source depends on the velocity of the source relative to the observer. The motion of the source causes the waves in front of the source to be compressed and those behind the source to be stretched (see Figure 1.1).

The entry of Christian Doppler in the *Biographical Encyclopedia of Scientists* describes an experimental test on the Doppler effect with sound waves: "Doppler's principle was tested experimentally in 1843 by Christoph Buys Ballot, who used a train to pull trumpeters at different speeds past musicians who had perfect pitch." The wavelength of the sound source is defined by $\lambda = c_{sound}/f$, where c_{sound} is the propagation speed of the sound wave in a given medium and f is the frequency of the sound source. If the source is moving at a velocity v_s relative to the medium, the frequency perceived by the observer is

$$f' = \frac{c_{sound}}{c_{sound} \mp v_s} f = \frac{1}{1 \mp v_s/c_{sound}} f \qquad (1.1)$$

If $v_s/c_{sound} \ll 1$, the Doppler shifted frequency perceived by the observer is approximately

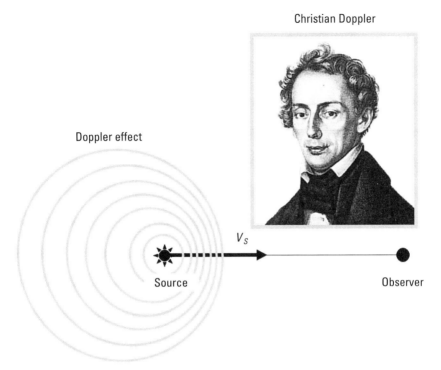

Christian Doppler

Doppler effect

V_S

Source

Observer

Figure 1.1 In 1842, Christian Doppler first discovered the phenomenon on the apparent color of a light source changed by its motion, known as the Doppler effect.

$$f' = \frac{1}{1 \mp v_s/c_{sound}} f \cong \left(1 \pm \frac{v_s}{c_{sound}}\right) f \qquad (1.2)$$

If the source is stationary and the observer is moving with a velocity v_0 relative to the medium, the frequency perceived by the observer becomes

$$f' = \frac{c_{sound} \pm v_o}{c_{sound}} f = \left(1 \pm \frac{v_o}{c_{sound}}\right) f \qquad (1.3)$$

If both the source and the observer are moving, the frequency perceived by the observer becomes

$$f' = \frac{c_{sound} \pm v_o}{c_{sound} \mp v_s} f = \frac{1 \pm v_o/c_{sound}}{1 \mp v_s/c_{sound}} f \qquad (1.4)$$

When the source and the observer moves toward each other, the upper set of signs in (1.4) is applied; when the source and the observer moves away from each other, the lower set of signs is applied.

1.2 Relativistic Doppler Effect and Time Dilation

Compared to sound waves, there is no medium involved in light or EM wave propagation. The propagation speed of light or EM waves, c, viewed from both the source and the observer, is the same constant.

For light or EM waves, changes in the frequency or wavelength caused by the relative motion between the source and the observer should take into account the effect of the theory of special relativity [6]. Thus, the Doppler frequency shift must be modified to be consistent with the Lorentz transformation. The relativistic Doppler effect is different from the classical Doppler effect because it includes the time dilation effect of special relativity and does not involve the medium of the wave propagation as a reference point.

When a light or EM source at a frequency, f, is moving with a velocity v_s at an angle θ_s relative to the direction from the source S to the observer O as shown in Figure 1.2, the time interval between two successive crests of the wave emitted at t_1 and t_2 is determined by

$$\Delta t_s = t_2 - t_1 = \frac{\gamma}{f} \tag{1.5}$$

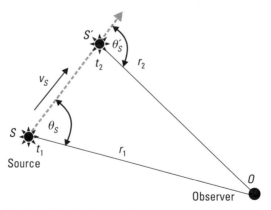

Figure 1.2 The Doppler effect in the case that only the source S is moving with a velocity v_s at an angle θ_s relative to the direction from the source S to the observer O.

where $\gamma = 1/(1-v_s^2/c^2)^{1/2}$ is a factor that represents the relativistic time dilation and c is the propagation speed of the light or EM wave. Then the time interval between arrivals of the two successive wave crests at the observer is

$$\Delta t_o = \left(t_2 + \frac{r_2}{c}\right) - \left(t_1 + \frac{r_1}{c}\right) = \frac{\gamma}{f}\left(1 - \frac{v_s \cdot \cos\theta_s}{c}\right) \tag{1.6}$$

Thus, the corresponding observed frequency by the observer becomes

$$f' = \frac{1}{\Delta t_0} = \frac{1}{\gamma}\frac{f}{1 - \dfrac{v_s \cos\theta_s}{c}} \tag{1.7}$$

If the angle, θ_s', between the moving direction of the source and the direction from the source to the observer is measured at the time when the wave arrives at the observer, the observed frequency by the observer becomes

$$f' = \gamma\left(1 + \frac{v_s \cos\theta_s'}{c}\right)f \tag{1.8}$$

Thus, the two angles θ_s and θ_s' are related by

$$\cos\theta_s = \frac{\cos\theta_s' + v_s/c}{1 + \dfrac{v_s \cos\theta_s'}{c}} \tag{1.9}$$

or

$$\cos\theta_s' = \frac{\cos\theta_s - v_s/c}{1 - \dfrac{v_s \cos\theta_s}{c}} \tag{1.10}$$

If both the source and the observer move as illustrated in Figure 1.3 in the two-dimensional (2-D) case, the observed frequency at the time when the wave is emitted is similar to (1.4) as

$$f' = \frac{1}{\gamma}\frac{1 \pm \dfrac{v_o \cos\theta_o}{c}}{1 \mp \dfrac{v_s \cos\theta_s}{c}}f \tag{1.11}$$

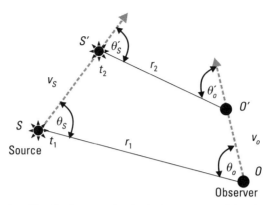

Figure 1.3 Doppler effect in the case that both the source and the observer move.

where θ_s and θ_o, as illustrated in Figure 1.3, are the angle of the source motion and the angle of the observer motion at the time when the wave is emitted, respectively.

In general, given the relative motion between the source and the observer v, when the source and the observer move toward each other, the observed Doppler shifted frequency can be rewritten as

$$f' = \frac{1}{\gamma} \frac{f}{1 - v/c} = \sqrt{1 - \left(\frac{v}{c}\right)^2} \frac{f}{1 - v/c} = \sqrt{\frac{1 + v/c}{1 - v/c}} f \qquad (1.12)$$

If the source is moving away from the observer, the observed frequency becomes

$$f' = \sqrt{\frac{1 - v/c}{1 + v/c}} f \qquad (1.13)$$

If the velocity v is much lower than the velocity of the EM wave propagation c, that is, $v \ll c$ or $v/c \approx 0$, the relativistic Doppler frequency is the same as the classical Doppler frequency.

According to the MacLaurin series,

$$\sqrt{\frac{1 - v/c}{1 + v/c}} = 1 - \frac{v}{c} + \frac{(v/c)^2}{2} - \cdots \qquad (1.14)$$

when the source and the observer are moving away from each other, the Doppler shifted frequency can be approximated by

$$f' \cong \left(1 - \frac{v}{c}\right)f \tag{1.15}$$

This is the same as the classical Doppler frequency shift. Thus, the Doppler frequency shift between the emitted frequency of the source and the perceived frequency by the observer is

$$f_D \cong f' - f = -\frac{v}{c}f \tag{1.16}$$

The Doppler frequency shift is proportional to the emitted frequency f of the wave source and the relative velocity v between the source and the observer.

1.3 Doppler Effect Observed in Radar

In radar, the velocity of a target, v, is usually much slower than the speed of the EM wave propagation c, that is, $v << c$ or $\beta = v/c \approx 0$. In monostatic radar systems, where the wave source (radar transmitter) and the receiver are at the same location, the round-trip distance traveled by the EM wave is twice the distance between the transmitter and the target. In this case, the wave movement consists of two segments: traveling from the transmitter to the target that produces a Doppler shift ($-fv/c$), and traveling from the target back to the receiver that produces another Doppler shift ($-fv/c$), where f is the transmitted frequency. Thus, the total Doppler shift becomes

$$f_D = -f\frac{2v}{c} \tag{1.17}$$

If the radar is stationary, v will be the radial velocity of the target along the LOS of the radar. Velocity is defined to be positive when the object is moving away from the radar. As a consequence, the Doppler shift becomes negative.

In a bistatic radar system as shown in the 2-D case in Figure 1.4, the transmitter and receiver are separated by a baseline distance L, which is comparable with the maximum range of a target with respect to the transmitter and the receiver. The range from the transmitter to the target is given by a vector, r_T, and the range from the receiver to the target is given by a vector, r_R, where a boldface letter is used to denote the vector quantity. The bistatic angle β is defined by the angle between the transmitter-to-target line and the receiver-to-target line. The transmitter look angle is α_T and the receiver look angle is α_R as illustrated in Figure 1.4. The look angle is defined by the

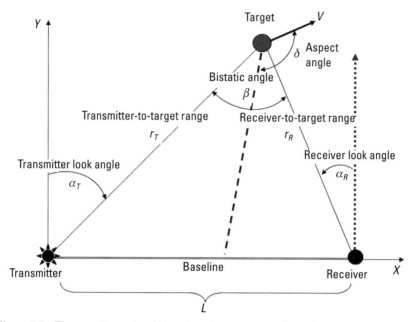

Figure 1.4   The two-dimensional bistatic radar system configuration.

angle from a corresponding reference vector perpendicular to the transmitter-receiver baseline to the target's LOS vector. The positive angle is defined in a counterclockwise direction. Thus, the bistatic angle $\beta = \alpha_R - \alpha_T$.

If the distance from the transmitter to the target is known by $r_T = |\mathbf{r}_T|$, the distance from the receiver to the target is

$$r_R = |\mathbf{r}_R| = \left(L^2 + r_T^2 - 2r_T L \sin\alpha_T \right)^{1/2} \tag{1.18}$$

and the receiver look angle becomes

$$\alpha_R = \tan^{-1}\left(\frac{L - r_T \sin\alpha_T}{r_T \cos\alpha_T} \right) \tag{1.19}$$

When the target is moving with a velocity vector $\mathbf{V}$, the component along the LOS direction from the transmitter to the target is

$$v_T = V \frac{\mathbf{r}_T}{|\mathbf{r}_T|} \tag{1.20}$$

and the component along the LOS direction from the receiver to the target is

$$v_R = V \frac{r_R}{|r_R|} \tag{1.21}$$

Then, due to the target's motion, the range from the transmitter to the target is a function of time

$$r_T(t) = r_T(t=0) + v_T t \tag{1.22}$$

and the range from the receiver to the target is also a function of time

$$r_R(t) = r_R(t=0) + v_R t \tag{1.23}$$

The phase change between the transmitted and the received signal is a function of the radar wavelength $\lambda = c/f$, the distance from the transmitter to the target $r_T(t)$, and the distance from the target to the receiver $r_R(t)$:

$$\Delta\Phi(t) = \frac{r_T(t) + r_R(t)}{\lambda} \tag{1.24}$$

Then the Doppler frequency shift is measured by the phase change rate. By taking the time derivative of the phase change, the bistatic Doppler frequency shift is

$$f_{D_{Bi}} = \frac{1}{2\pi} \frac{d}{dt} \Delta\Phi(t) = \frac{1}{2\pi} \frac{1}{\lambda} \left[\frac{d}{dt} r_T(t) + \frac{d}{dt} r_R(t) \right] = \frac{1}{2\pi} \frac{1}{\lambda} (v_T + v_R) \tag{1.25}$$

To track the change of phase with time, the phase of the transmitted signal must be exactly known. Thus, a fully coherent system is required to preserve and track the phase change in the received signal.

In a bistatic radar system, the Doppler shift depends on three factors [7]. The first factor is the maximum Doppler shift. If a target is moving with a velocity V, the maximum Doppler shift is

$$f_{D_{max}} = \frac{2f}{c} |V| \tag{1.26}$$

The second factor is related to the bistatic triangulation factor:

$$D = \cos\left(\frac{\alpha_R - \alpha_T}{2}\right) = \cos\left(\frac{\beta}{2}\right) \tag{1.27}$$

The third factor is related to the angle δ between the direction of the target's moving and the direction of the bisector: $C = \cos\delta$.

Thus, the Doppler shift of a bistatic radar system can be represented by

$$f_{D_{Bi}} = f_{D_{max}} D \cdot C = \frac{2f}{c}|V|\cos\left(\frac{\beta}{2}\right)\cos\delta \qquad (1.28)$$

If two targets are separated in range and velocity, these targets can be resolved by the range resolution and Doppler resolution of the radar system. In a monostatic radar system, if the range resolution is known as Δr_{Mono} and the Doppler resolution is known as Δf_{DMono}, then the range resolution and the Doppler resolution of a bistatic radar system can be determined by the corresponding monostatic range resolution and Doppler resolution scaled by a function of the bistatic angle β. Thus, the bistatic range resolution is

$$\Delta r_{Bi} = \frac{1}{\cos(\beta/2)}\Delta r_{Mono} \qquad (1.29)$$

and the bistatic Doppler resolution is

$$\Delta f_{D_{Bi}} = \cos\left(\frac{\beta}{2}\right)\Delta f_{D_{Mono}} \qquad (1.30)$$

However, in an extreme case where the bistatic angle is near 180°, the radar becomes a forward-scattering radar [8]. The EM field scattered in the forward direction is 180° out of phase with the incident field. Thus, it removes power from the incident field and forms a shadow area behind the target. Based on (1.28), in the forward scattering case, the bistatic Doppler frequency shift becomes zero regardless the actual target velocities. This can also be explained by the fact that as the target crosses the baseline, the transmitter-to-target range changes in an equal and opposite way to the target-to-receiver range; thus, the Doppler shift must be zero.

1.4 Estimation and Analysis of Doppler Frequency Shifts

Doppler radars utilize the Doppler effect to measure the radial velocity of a moving target. The Doppler frequency shift can be extracted by a quadrature detector that produces an in-phase (I) component and a quadrature phase (Q) component from the input signal as shown in Figure 1.5.

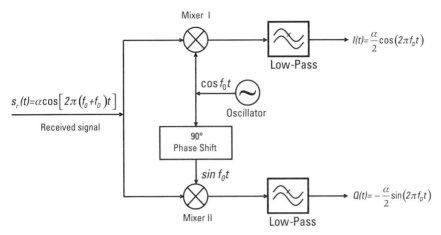

Figure 1.5 Doppler shifts extracted by a quadrature detector.

In the quadrature detector, the received signal is split into two mixers called synchronous detectors. In the synchronous detector I, the received signal is mixed with a reference signal, the transmitted signal; in the other channel, it is mixed with a 90° shift of the transmitted signal.

If the received signal is expressed as

$$s_r(t) = a\cos\left[2\pi\left(f_0 + f_D\right)t\right] = a\cos\left[2\pi f_0 t + \varphi(t)\right] \qquad (1.31)$$

where a is the amplitude of the received signal, f_0 is the carrier frequency of the transmitter, and $\varphi(t) = 2\pi f_D t$ is the phase shift on the received signal due to the target's motion. By mixing with the transmitted signal

$$s_t(t) = \cos\left(2\pi f_0 t\right) \qquad (1.32)$$

the output of the synchronous detector I is

$$s_r(t)s_t(t) = \frac{a}{2}\cos\left[4\pi f_0 t + \varphi(t)\right] + \frac{a}{2}\cos\varphi(t) \qquad (1.33)$$

After lowpass filtering, the I-channel output is

$$I(t) = \frac{a}{2}\cos\varphi(t) \qquad (1.34)$$

By mixing with the 90° phase-shifted transmitted signal

$$s_t^{90°}(t) = \sin\left(2\pi f_0 t\right) \tag{1.35}$$

the output of the synchronous detector II is

$$s_r(t)s_t^{90°}(t) = \frac{a}{2}\sin\left[4\pi f_0 t + \varphi(t)\right] - \frac{a}{2}\sin\varphi(t) \tag{1.36}$$

After lowpass filtering, the Q-channel output is

$$Q(t) = -\frac{a}{2}\sin\varphi(t) \tag{1.37}$$

Combining the I & Q outputs, a complex Doppler signal can be formed by

$$s_D(t) = I(t) + jQ(t) = \frac{a}{2}\exp\left[-j\varphi(t)\right] = \frac{a}{2}\exp\left(-j2\pi f_D t\right) \tag{1.38}$$

Thus, the Doppler frequency shift f_D can be estimated from the complex Doppler signal $s_D(t)$ by using a frequency measurement tool.

To estimate the Doppler frequency shift of a single sinusoidal signal, the periodogram can be used to calculate the spectral density of the signal. Then the maximum likelihood estimation can be applied to locate the maximum of the periodogram [9, 10]

$$\hat{f}_D = \max_{f_D(k)}\left\{\left|\sum_{k=1}^{N} a(k)\exp\left(-j2\pi f_D(k)\right)\right|^2\right\} \tag{1.39}$$

When the number of samples in the analyzed signal is limited, to estimate the signal spectral, the simplest method is to use the fast Fourier transform (FFT), which is computationally efficient and easy to implement. However, its frequency resolution is limited to the reciprocal of the time interval of the signal and suffers from spectrum leakage associated with the time windowing. Manually increasing the time window with zero padding corresponds to a higher interpolation density in the frequency domain, but not a higher-frequency resolution. The usual way to increase frequency resolution is to take FFT with a longer time duration of the analyzed signal without zero padding. However, the computation time of the FFT is of the order of $O(N \times \log N)$, where N is the number of samples in the analyzed signal. For large number of samples N, the FFT is not computationally efficient.

To alleviate the limitations of the FFT, alternative spectral estimation methods were proposed [10, 11]. Autoregressive (AR) modeling and eigen-vector-based methods, such as the multiple signal classification (MUSIC) and other super-resolution methods for spectral analysis, can be used in the frequency estimation. However, they require either intensive matrix computations or iterative optimization techniques.

Because the frequency is determined by the time derivative of the phase function, the phase difference $\varphi(t)$ between the received and the transmitted signal can be used to calculate the instantaneous Doppler frequency shift f_D of the received signal

$$f_D = \frac{1}{2\pi} \frac{d\varphi(t)}{dt} \tag{1.40}$$

However, the instantaneous frequency is only suitable for monocomponent or single-tone signals, but not suitable for those containing multiple components. To deal with a signal having multiple components, an approach that decomposes a multicomponent signal into multiple monocomponent signals may be used. Then the complete time-frequency distribution of the multicomponent signal can be derived by computing the instantaneous frequency for each of the monocomponent signals and combining these instantaneous frequencies of monocomponent signals together. Further discussion about the instantaneous frequency can be found in Section 1.8.

For a single tone signal in additive Gaussian noise, if the phase difference $\Delta\varphi(k)$, ($k = 1, \dots N - 1$), from one sample to the next can be tracked, a weighted linear combination of these phase differences may be used to estimate the sinusoidal frequency tone [12]

$$\hat{f}_D = \frac{1}{2\pi} \sum_{k=1}^{N-1} w(k)\Delta\varphi(k) \tag{1.41}$$

where the weighting function $w(k)$ is

$$w(k) = \frac{6k(N-k)}{N(N^2-1)} \tag{1.42}$$

This estimation shows that at high signal-to-noise ratio (SNR), the frequency estimation attains the Cramer-Rao lower bound as described in the Section 1.5.

To preserve and track the phase of the received signals, the frequency source in the transmitter must keep a very high phase stability. Therefore, the radar must be fully coherent to keep an accurate phase coherency.

From the estimated Doppler frequency, the radial velocity of a target is determined by

$$v = \frac{\lambda}{2} \hat{f}_D = \frac{c}{2f} \hat{f}_D \tag{1.43}$$

The I&Q outputs of the quadrature detector can also be used to determine whether the target is approaching to or away from the radar. As illustrated in Figure 1.6, by comparing the relative phase between the I-channel and the 90°-shifted Q-channel, two flow channels (one is approaching the radar and the other is "away" from the radar) can be produced.

Doppler radars include pure continuous-wave (CW) radar without modulations, frequency-modulated continuous-wave (FMCW) radar, and coherent pulsed Doppler radar. Pure CW radars can only measure the velocity. FMCW and coherent pulsed Doppler radars can have wide frequency bandwidth to gain a high-range resolution and measure both the range and Doppler information. Coherent Doppler radars retain the phase of the transmitted signals and track the phase changes in the received signals. Doppler frequency shift is proportional to the phase change rate. If the phase change is more than $\pm\pi$, the estimated Doppler frequency becomes ambiguous called

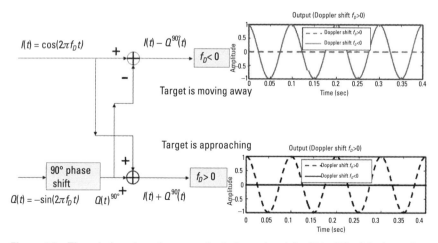

Figure 1.6 The relative phase between the I-channel and the 90°-shifted Q-channel to determine whether the target is approaching or going away from the radar.

the Doppler aliasing. This is caused by discrete time sampling of a continuous time signal. The sampling process can be represented by the multiplication of the continuous time signal $s(t)$ with a sequence of delta functions $\delta(t)$. Taking the Fourier transform of the time-sampled signal, the discrete time-sampled signal is transformed to the frequency domain represented by a discrete form of the Fourier transform:

$$s(t) \times \sum_n \delta(t - n\Delta t) \Rightarrow S(f) \otimes \sum_m \delta(f - m/\Delta t) \qquad (1.44)$$

where Δt is the time-sampling interval, the convolution operator $\otimes$ in the frequency domain makes the signal spectrum $S(f)$ replicated at a period of $1/\Delta t$. If the frequency bandwidth of the signal spectrum is greater than the Nyquist frequency $1/(2\Delta t)$, or the sampling rate is lower than the half-bandwidth of the signal, this replication will cause signal spectrum to overlap and produce ambiguity, called aliasing.

Figure 1.7 illustrates the aliasing phenomenon. The frequency-modulated signal is sampled with a sampling rate lower than the Nyquist rate $\pm 1/(2\Delta t)$. Spectrum aliasing can be seen clearly in Figure 1.7(a). The time-varying frequency spectrum of the signal is shown in Figure 1.7(c), where modulated frequency values in excess of the Nyquist rate $\pm 1/(2\Delta t)$ are aliasing. The aliasing causes the true frequency values to be offset by multiples of $(1/\Delta t)$ until they fall into the Nyquist cointerval. For example, if the Nyquist frequency is $\pm 1,000$ Hz, a frequency value of $+1,500$ Hz is aliased to $+1,500$ Hz $- 2 \times 1,000$ Hz $= -500$ Hz and $-1,500$ Hz is aliased to $-1,500$ Hz $+ 2 \times 1,000$ Hz $= +500$ Hz. In this case, the true frequency values are offset by one multiple of $(1/\Delta t)$ to fall into the Nyquist interval as illustrated in Figure 1.7(c).

To resolve the aliasing ambiguity, increasing the sampling rate or techniques that interpolate missing data points may be applied. Figure 1.7(b) is the spectrum of the same signal sampled with twice the original sampling rate. The time-varying spectrum with twice the original sampling rates is shown in Figure 1.7(d), where the full time-varying spectrum is restored.

Generally, the required unambiguous radial velocity must be at least the radial velocity with which the target moves. Thus, the unambiguous velocity that a radar can be measured depends on the transmitted frequency f or wavelength $\lambda = c/f$ and the time interval Δt between two sampling points

$$v_{\max} = \frac{f_{D\max}}{2f/c} = \pm \frac{\lambda}{4\Delta t} \qquad (1.45)$$

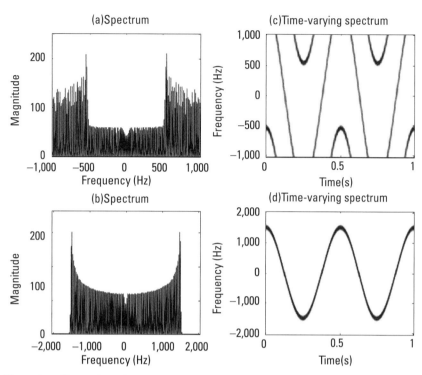

Figure 1.7   Illustration of the aliasing phenomenon. (a) Spectrum with aliasing. (b) The spectrum of the same signal but sampled with twice the original sampling rate. (c) The time-varying spectrum of the signal used in (a). (d) The time-varying spectrum of the signal used in (b).

For a coherent pulsed radar, which measures both the velocity and range, the time interval Δt is equal to $1/PRF$, where the PRF is the pulse repetition frequency. Thus, the maximum velocity that the pulsed radar can measure unambiguously is

$$v_{max} = \pm\frac{\lambda PRF}{4} \tag{1.46}$$

Velocities greater than $\lambda\ PRF/4$ are folded into $\pm\lambda\ PRF/4$ called the Nyquist velocity.

The range that can be measured by the pulsed radar is limited by the maximum unambiguous range

$$r_{max} = \frac{c}{2PRF} \tag{1.47}$$

Ranges greater than the r_{max} are folded into the first range region. Without additional information, the correct range information cannot be determined.

The pulse repetition frequency (PRF) is proportional to the maximum unambiguous velocity, but inversely proportional to the maximum range being measured. A compromise between the maximum range and maximum velocity is always desirable. The maximum velocity can be extended by using two alternating PRFs. Since the maximum velocity is related to the wavelength, a longer wavelength or a lower frequency can increase the limit of the maximum velocity.

However, the product of the maximum unambiguous velocity and the unambiguous range

$$v_{max} r_{max} = \pm \frac{c\lambda}{8} = \pm \frac{c^2}{8f} \qquad (1.48)$$

is not directly related by the PRF. It is only determined by the frequency f or the wavelength λ. Given a frequency band, the product of the maximum unambiguous velocity and the unambiguous range is a constant. Increasing the maximum unambiguous velocity will decrease the maximum unambiguous range and vice versa. The trade-off between the unambiguous velocity and unambiguous range is often referred to as the *Doppler dilemma*.

In pulsed radars, to avoid aliasing, very high PRF should be selected; to avoid ambiguous range, a very low PRF is required. However, the low PRF also limits the extracted Doppler information. To have a suitable range ambiguity and aliasing, multiple PRFs are often used.

1.5 Cramer-Rao Bound of the Doppler Frequency Estimation

In practice, the Doppler frequency estimation is considered in the presence of noise. From the estimation theory, to estimate the value of an unknown parameter θ from N noisy measurements, if the expected value of the estimate equals the true value of the parameter $E\{\hat{\theta}\} = 0$, the estimator is said to be unbiased. Otherwise, the estimator is biased. When the estimator asymptotically converges in probability $\Pr\{\cdot\}$ to the true value (i.e., $\lim_{N\to\infty} \Pr\{|\hat{\theta} - \theta| > \varepsilon\} = 0$, where ε is an arbitrary small positive number), this is a consistent estimator.

The benchmark that evaluates the variance of a particular unbiased estimator can be described by the Cramer-Rao lower bound (CRLB) that provides a lower bound on the variance of a linear or nonlinear unbiased estimator and gives an insight into the performance of the estimator [13, 14]. It states that the variance of an unbiased estimator is at least as high as the inverse of the Fisher information.

If an unknown deterministic parameter θ is estimated from the N statistical measurements x_k, ($k = 1, \ldots, N$), with the probability density function of $p(x_k; \theta)$, the variance of the unbiased estimation $var\{\hat{\theta}\}$ is bounded by the inverse of the Fisher information $I(\theta)$, that is,

$$var\{\hat{\theta}\} \geq \frac{1}{I(\theta)} \tag{1.49}$$

The Fisher information is defined by

$$I(\theta) = E\left\{\left[\frac{\partial}{\partial \theta} \log p\left(x_k; \theta\right)\right]^2\right\} = -E\left\{\frac{\partial^2}{\partial \theta^2} p\left(x_k; \theta\right)\right\} \tag{1.50}$$

where $E\{\cdot\}$ means the expectation value that is taken with respect to $p(x_k; \theta)$ and results in a function of θ.

To estimate a single sinusoidal Doppler frequency in white Gaussian noise, the Fisher information can be inverted easily. The CRLB of the Doppler frequency estimation can be derived as

$$var\left(\hat{f}_D\right) \geq \frac{6}{N\left(N^2 - 1\right) \cdot SNR} \tag{1.51}$$

where SNR is the signal-to-noise ratio and N is the number of samples of the signal [14].

1.6 The Micro-Doppler Effect

The micro-Doppler effect was originally introduced in coherent *laser* (*light amplification by stimulated emission of radiation*) radar systems [5]. A laser detection and ranging (LADAR) system transmits an EM wave at optical frequencies to an object, and receives the reflected or backscattered light wave to measure the object's range, velocity, and other properties through modulations of its laser beam by amplitude, frequency, phase, and even polarization.

Coherent LADAR, which preserves the phase information of the scattered light wave with respect to a reference laser wave generated in the local oscillator, has greater sensitivity to phase changing and is capable of measuring object velocity from the phase change rate.

In a coherent system, because the phase of a returned signal from an object is sensitive to the variation in range, a half-wavelength change in range causes 360° phase change. For LADAR with a wavelength of 2 μm, a 1-μm range variation causes a 360° phase change. In the case of vibration, if the vibration frequency is f_v and the amplitude of the vibration is D_v, the maximum Doppler frequency variation is determined by

$$\max\{f_D\} = \left(\frac{2}{\lambda}\right)D_v f_v \qquad (1.52)$$

As a consequence, in a high-frequency system, even with very low vibration rate f_v, a very small vibration amplitude D_v can cause a large phase change, and, thus, Doppler frequency shifts can be easily detected.

In many cases, an object or any structural component of the object may have micromotions in addition to the bulk motion of the object (including zero bulk motion). The term *micromotion* includes a broader usage of the *micro*, such that any small motion (such as vibration, oscillation, rotation, swinging, flapping, and even fluctuation) can be called as the micromotion. The source of micromotion may be a vibrating surface, the rotating rotor blades of a helicopter, a walking person with swinging arms and legs, the flapping wings of a bird, or other causes.

Human motion is an important topic in micro-Doppler research. Human articulated motion is accomplished by a series of motion of human body parts. It is a complex micromotion due to high articulation and flexibility. Walking is a typical example of human articulated motion.

Micromotion induces frequency modulations on the carrier frequency of radar transmitted signals. For a pure periodic vibration or rotation, micro motion generates side-band Doppler frequency shifts about the center of the Doppler shifted carrier frequency. The modulation contains harmonic frequencies determined by the carrier frequency, the vibration or rotation rate, and the angle between the direction of vibration and the direction of the incident wave. The frequency modulation enables us to determine kinematic properties of the object of interest. The time-varying frequency modulation can serve as a signature of the object for further classification, recognition, and identification.

1.7 Micro-Doppler Effect Observed in Radar

The micro-Doppler effect is sensitive to the frequency band of the signal. For a radar system operating at a microwave frequency band, the micro-Doppler effect of a vibrating target may be observable if the product of the vibration rate and the displacement of the vibration is high enough. For example, a radar operating at the X-band with a 3-cm wavelength, a vibration rate of 15 Hz with a displacement of 0.3 cm can induce a detectable maximum micro-Doppler frequency shift of 18.8 Hz. If the radar operates at the L-band with 10-cm wavelength, to achieve the same maximum micro-Doppler shift of 18.8 Hz, for the same vibration rate of 15 Hz, the required displacement must be 1 cm, which may be too large to be achieved in practice. Therefore, in radar systems operating at lower-frequency bands, micro-Doppler shifts generated by the vibration may not be detectable. However, micro-Doppler shifts generated by rotations, such as rotating rotor blades, may be detectable because of their longer rotation arms and thus higher tip speeds.

The ultrahigh-frequency (UHF)-band radar operating at a frequency band of 300–1,000 MHz is widely used for foliage penetration (FOPEN) to detect targets under trees. In FOPEN radars, the micro-Doppler shift induced by a target's vibrations is usually too small to be detected. However, it is still possible to detect micro-Doppler shifts generated by rotating rotor blades or propellers. For a radar operating at the UHF-band with a 0.6-m wavelength, if a helicopter's rotor blade rotates with a tip speed of 200 m/s, its maximum micro-Doppler shift can reach 666 Hz and it is certainly detectable.

1.8 Estimation and Analysis of Micro-Doppler Frequency Shifts

The micro-Doppler shift is a time-varying frequency shift that can be extracted from the complex output signal of a quadrature detector used in the conventional Doppler radar. For analyzing time-varying frequency features, the Fourier transform is not suitable because it cannot provide time-dependent frequency information. The commonly used analysis methods to describe a signal simultaneously in the time and frequency domains are the instantaneous frequency analysis and the joint time-frequency analysis.

The terminology of the instantaneous frequency defined by the time derivative of the phase function in a time-varying signal has been argued decades ago because the amplitude and phase functions are not unique. A well-accepted instantaneous frequency definition uses a pair of Hilbert transform to form the real part and the imaginary part of an analytic signal [15]. Thus, the *instantaneous* term means in the sense of the present time instant, and its measurement requires only the knowledge of the analyzed signal over the past and not from the future.

The instantaneous frequency derived by the time-derivative operation yields only one value of frequency at a given time instant. This means that it is only suitable for monocomponent signals and not for multicomponent signals. A monocomponent signal is narrowband at any time and has energy in a contiguous portion in the joint time-frequency domain. Conversely, a multicomponent signal has energy in multiple isolated frequency bands at the same time instant. To deal with multicomponent signals, an obvious approach is to decompose the multicomponent signal into multiple addable monocomponent signal components [16]. The complete time-frequency distribution of the signal is obtained by computing the instantaneous frequencies for each component signal and combining these individual instantaneous frequencies together.

Joint time-frequency analysis has been used for decades for analyzing the time-varying frequency spectrum. It is designed to localize the energy distribution of a given signal in the 2-D time and frequency domains. It is quite suitable for not only monocomponent signals but also multicomponent signals.

1.8.1 Instantaneous Frequency Analysis

The instantaneous frequency is an important representation for nonstationary signal analysis. For a real valued signal $s(t)$, its associated complex valued signal $z(t)$ is defined by

$$z(t) = s(t) + jH\{s(t)\} = a(t)\exp\left[\varphi(t)\right] \tag{1.53}$$

where $H\{\cdot\}$ is the Hilbert transform of the signal given by

$$H\{s(t)\} = \frac{1}{\pi} \int_{-\infty}^{\infty} \frac{s(\tau)}{t-\tau} d\tau \tag{1.54}$$

$z(t)$ is called the analytic signal associated to $s(t)$, $a(t)$ is the amplitude function, and $\varphi(t)$ is the phase function of the analytic signal. In the frequency domain, the Fourier transform of the analytic signal, $Z(f)$, is single-sided with zero values at negative frequencies and double values at positive frequencies. Thus, the instantaneous frequency of the signal $z(t)$ is the time-derivative of the uniquely defined phase function $\varphi(t)$ of the analytic signal

$$f(t) = \frac{1}{2\pi} \frac{d}{dt} \varphi(t) \tag{1.55}$$

In practice, a discrete real-valued signal $s(n)$ and sampled at time instants $t = n\Delta t$, $n = 1, 2, \ldots, N$ may be used. Then, the discrete analytic signal $z(n)$ becomes

$$z(n) = s(n) + j\,\mathrm{H}\{s(n)\} \tag{1.56}$$

For a discrete signal, the instantaneous frequency is similar to (1.55), but with discrete derivatives of the phase, which can be estimated by using the central finite difference equation of the phase function [17]:

$$f(n) = \frac{1}{2\pi}\frac{1}{2\Delta t}\big[\varphi(n+1) - \varphi(n-1)\big]_{2\pi} \tag{1.57}$$

where Δt is the sampling interval and $[\cdot]_{2\pi}$ represents reduction modulo 2π, and n is the discrete number of time samples.

Instantaneous frequency only gives one value at a time and is only good for describing signals comprised of a single oscillating frequency component at a time, called the monocomponent signal. It is not suitable for signals having several different oscillating frequency components at a time (i.e., multicomponent signal). To distinguish frequency contributions of a multicomponent signal, it is necessary to preprocess the multicomponent signal into its monocomponent elements. Huang et al. [16] introduced the concept of empirical mode decomposition (EMD) to separate a multicomponent signal into monocomponent constituents by a progressive sifting process to yield the bases called the intrinsic mode functions (IMFs). Later, Olhede and Walden [18] introduced a wavelet packet-based decomposition as a replacement of the EMD in preprocessing the multicomponent signal.

The EMD adaptively decomposes a signal into a limited number of zero-mean, narrowband IMFs. Then the instantaneous frequency of each IMF is calculated by using the normalized Hilbert transform, called the Hilbert-Huang transform (HHT) [16]. The combination of the Hilbert spectrum is the complete time-varying frequency spectrum.

The original formulation of the EMD can only be applied to real-valued signals. However, in radar applications, signals are always complex with I and Q parts. The extension of the EMD to handle complex-valued signals has been proposed in [19, 20]. The detailed procedure of calculating complex EMD and HHT with MATLAB codes can be obtained from [21].

1.8.2 Joint Time-Frequency Analysis

When the spectral composition of a signal varies as a function of time, the conventional Fourier transform cannot provide a time-dependent spectral description. Thus, a joint time-frequency analysis provides more insight into the time-varying behavior of the signal.

Motivated by defining the information content in a signal, in 1946, Dennis Gabor, a Hungarian Nobel laureate, proposed the first algorithm on the time-frequency analysis of an arbitrary signal [22]. Gabor suggested that the time and frequency characteristics of a signal $s(t)$ can be simultaneously observed by using the expansion

$$s(t) = \sum_{m=-\infty}^{\infty} \sum_{n=-\infty}^{\infty} a_{mn} G(g,n,m) \tag{1.58}$$

where $G(g,n,m)$, called the Gabor function, is expressed in terms of a Gaussian window $g(t)$ by

$$G(g,n,m) = g(t - m\Delta T)e^{jn\Delta Ft} \tag{1.59}$$

where ΔT and ΔF are the time and frequency lattice intervals, respectively, and the Gaussian window is defined by

$$g(t) = \frac{1}{\pi^{1/4}\sqrt{\sigma}} \exp\left\{-\frac{t^2}{2\sigma^2}\right\} \tag{1.60}$$

Gabor claimed that the basis functions $G(\cdot)$ used in this time-frequency decomposition have the minimum area in the joint time-frequency plane.

The spectrogram is a widely used method to display time-varying spectral density of a time-varying signal. It is a spectro-temporal representation and provides the actual change of frequency contents of a signal over time. The spectrogram is calculated by using the short-time Fourier transform (STFT) and represented by the squared magnitude of the STFT without keeping phase information of the signal

$$\text{Spectrogram}(t,f) = \left|STFT(t,f)\right|^2 \tag{1.61}$$

The STFT performs the Fourier transform on a short-time window basis rather than taking the Fourier transform to the entire signal using one long-time window.

With the time-limited window function, the resolution of the STFT is determined by the window size. There is a trade-off between the time resolution and the frequency resolution. A larger window size has a higher frequency resolution but a poorer time resolution. The Gabor transform is a typical short-time Fourier transform using Gaussian windowing and has the minimal product of the time resolution and the frequency resolution.

To better analyze the time-varying micro-Doppler frequency characteristics and visualize the localized joint time and frequency information, the signal must be analyzed by using a high-resolution time-frequency transform to characterize the spectral and temporal behavior of the signal. Bilinear transforms, such as the Wigner-Ville distribution (WVD), are high-resolution time-frequency transforms. The WVD of a signal $s(t)$ is defined by the Fourier transform of the time-dependent auto-correlation function

$$\text{WVD}(t,f) = \int s\left(t+\frac{t'}{2}\right)s^*\left(t-\frac{t'}{2}\right)\exp\{-j2\pi ft'\}dt' \qquad (1.62)$$

where $s(t+t'/2)s^*(t-t'/2)$ can be interpreted as a time-dependent autocorrelation function. The bilinear WVD has a better joint time-frequency resolution than any linear transform, such as the STFT. However, it suffers from the problem of cross-term interference (i.e., the WVD of the sum of two signals is not the sum of their individual WVDs). If a signal contains more than one component in the joint time-frequency domain, its WVD will contain cross-terms that occur halfway between each pair of autoterms. The magnitude of these oscillatory cross-terms can be twice as large as the autoterms. To reduce the cross-term interference, filtered WVDs have been used to preserve the useful properties of the time-frequency transform with a slightly reduced time-frequency resolution and a largely reduced cross-term interference. The WVD with a linear lowpass filter belongs to the Cohen class [23].

The general form of Cohen class is defined by

$$C(t,f) = \iint s\left(u+\frac{\tau}{2}\right)s^*\left(u-\frac{\tau}{2}\right)\phi(t-u,\tau)\exp\{-j2\pi f\tau\}du\,d\tau \quad (1.63)$$

The Fourier transform of the lowpass filter $\phi(t,\tau)$, denoted as $\Phi(\theta,\tau)$, is called the kernel function. If $\Phi(\theta,\tau) = 1$, then $\phi(t,\tau) = \delta(t)$ and the Cohen class reduces to the WVD. The Cohen class with different kernel functions, such as the pseudo Wigner, the smoothed pseudo Wigner-Ville (SPWV), the Choi-Williams distribution, and the cone kernel distribution, can be used to largely reduce the cross-term interference in the WVD.

Other useful high-resolution time-frequency transforms are the adaptive Gabor representation and the time-frequency distribution series [24]. They decompose a signal into a family of basis functions, such as the Gabor function, which is well localized in both the time and the frequency domain and adaptive to match the local behavior of the analyzed signal.

In contrast with the EMD method, the adaptive Gabor representation is a signal-adaptive decomposition. It decomposes a signal $s(t)$ into Gabor basis functions $h_p(t)$ with an adjustable standard deviation σ_p and a time-frequency center (t_p, f_p):

$$s(t) = \sum_{p=1}^{\infty} B_p h_p(t) \tag{1.64}$$

where

$$h_p(t) = \left(\pi \sigma_p^2 \right)^{-1/4} \exp \left[-\frac{\left(t - t_p \right)^2}{2\sigma_p^2} \right] \exp \left(j 2\pi f_p t \right) \tag{1.65}$$

The coefficients B_p are found by an iterative procedure beginning with the stage $p = 1$ and choosing the parameters s_p, t_p, and f_p such that $h_p(t)$ is most similar to $s(t)$:

$$|B_p|^2 = \max_{\sigma_p, t_p, f_p} \left| \int s_{p-1}(t) h_p^*(t) \, dt \right|^2 \tag{1.66}$$

where $s_0(t) = s(t)$, that is, the analyzed signal is taken as the initial signal for $p = 1$. For $p > 1$, $s_p(t)$ is the residual after the orthogonal projection of $s_{p-1}(t)$ onto $h_p(t)$ has been removed from the signal:

$$s_p(t) = s_{p-1}(t) - B_p(t) h_p(t) \tag{1.67}$$

This procedure is iterated to generate as many coefficients as needed to accurately represent the original signal. Finally, the time-dependent spectrum is obtained by

$$\text{Adaptive Gabor} \, (f, t) = \sum_p |B_p|^2 \text{WVD}_{h_p}(t, f) \tag{1.68}$$

The well-known MATLAB time-frequency toolbox developed at the CNRS (Centre National de la Recherche Scientique) in France is a collection of time-frequency analysis tools [25]. It includes many commonly used linear and bilinear time-frequency distributions and can be used to compute micro-Doppler signatures represented in the joint time-frequency domain. However, this time-frequency toolbox was designed for analytic signals. Thus, rescaling the frequency scale in the output time-frequency representations is needed if the input signal is a complex I and Q signal.

1.9   The Micro-Doppler Signature of Objects

The term "signature" is commonly used to refer to the characteristic expression of an object or a process. For example, the characteristic mode in different ocean basins is called a signature of climate phenomenon, such as ENSO and El Niño. In Doppler weather radars, a special pattern of strong outbound and inbound winds constitutes the signature of a tornado.

When examining the Doppler phenomenon of an object, distinctive micro-Doppler characteristics provide evidence of the identity of the object's movement. The micro-Doppler signature is the distinctive characteristics of the movement. It is an intricate frequency modulation represented in the joint time and Doppler frequency domain, and it is distinctive characteristics that give an object its identity.

Figure 1.8 shows the micro-Doppler signature of a rotating air-launched cruise missile (ALCM) provided in the simulation software in [26, 27]. The length of the cruise missile is 6.4m and its wingspan is about 3.4m. A burst of a 1-μs chirp pulse radar operating in the X-band is assumed to simulate the EM backscattering field. The radar transmits 8,192 pulses with the pulse repetition interval of 67 μs during a period of 0.55 second to cover the total target's rotation angle of 360°. The simulated radar I and Q data from the rotating ALCM is provided in the companion MATLAB micro-Doppler signature analysis tools of the book.

The micro-Doppler features of the rotating ALCM can be observed in the frequency domain and with a much clearer view in the joint time-frequency domain [28]. Figure 1.8(b) shows the joint time-frequency micro-Doppler signature of the rotating ALCM. For comparison, the conventional Fourier spectrum is shown in Figure 1.8(a). Recall that the missile rotation rate is about 1.8 cycles/second because it takes 0.55 second to complete a rotation of 360°. For the ALCM model, (1) the missile head tip, (2) head joint, (3) wing joint, (4) turbine engine intake, (5) tail fin and tail plane, and (6) tail tip and engine exhaust are located at about −2.5m, −1.8m, 0.2m, 2.5m, 3.5m, and 4.2m, respectively, from the pivot point at 0. If these parts are considered to be the dominant scatterers, the maximum Doppler shifts induced by rotations of these parts would appear at those positions when their angular velocities are nearly in parallel with the radar LOS, when the missile is at 90° or 270° aspect to the radar, or at the elapsed time of 0.14 second or 0.41 second, respectively,

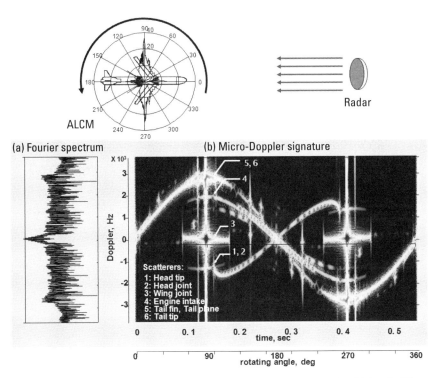

Figure 1.8 The micro-Doppler signature of a simulated rotating ALCM. (*After*: [28].)

as shown in Figure 1.8(b). When the missile is at a 90° aspect, the induced Doppler shifts are −1,917 Hz, −1,380 Hz, 153 Hz, 1,917 Hz, 2,684 Hz, and 3,217 Hz, respectively. For these dominant scatterers, the traces of induced Doppler shifts by rotations are clearly shown in Figure 1.8(b). At the aspects from 180° to 90° and from 180° to 270°, the induced Doppler shifts of the ALCM model have the same magnitudes but with opposite signs. The same motion kinematics is also seen at the missile aspects from 0° to 90° and from 360° to 270°. These Doppler shifts gradually reduce to zero when their moving directions are perpendicular to the radar LOS (i.e., when the missiles are approximately at 180° and 360° aspects). Notice that the induced Doppler shifts are further dispersed for two scatterers near the missile tail where two turbines are located. Thus, the missile kinematic motion can be well characterized by its micro-Doppler signature, which can be used to identify distinctive target features. The MATLAB source code for calculating the micro-Doppler signature of a rotating ALCM is provided in the book and can be used as an example for micro-Doppler signature analysis.

1.10 Angular Velocity Induced Interferometric Frequency Shift

The angular velocity describes the angular speed of an object in an angular motion about some axis, such as a rotating blade of a helicopter or a rolling wheel. The rate of change in angular displacement is known as the angular velocity. It is defined by radians per second or revolutions per second (rps). If an object is moving along a curved path, the velocity of the object is determined by both the rate of change of positional vector and directional change.

The instantaneous linear velocity of an object is called the tangential velocity. For an object moving along a circular path with radius r and angular velocity Ω (rad/sec), its tangential velocity is equal to the product of radius and angular velocity: $V_t = r\Omega$.

The angular velocity is a vector quantity, consisting of an angular speed and its direction. The magnitude of the angular velocity vector is directly proportional to the angular speed. The direction of the angular velocity vector is perpendicular to the plane in which the rotation takes place. If the rotation appears clockwise with respect to the observer, the angular velocity vector points away from the observer. If the rotation appears counterclockwise, the angular velocity vector points toward the observer.

Doppler radar is well known for measuring the radial velocity of a moving object. However, when the moving object has only angular velocity without the radial component, the radar is unable to measure true velocity of the object. If an object moving along a curved path, when its radial velocity decreases, the angular velocity must increase. Thus, for completely describing the object's motion, its angular velocity should be measured.

A radar technique of measuring the angular velocity of an object was proposed by J. Nanzer in [29–31]. An interferometric correlation receiver was proposed, where the receiver response frequency is proportional to the angular velocity. The method of correlation interferometer has long been used in the radio astronomy [32, 33].

A typical interferometric radar receiver has two separated receiver-channels with two antennas separated by a baseline D for observing a far-field source, as shown in Figure 1.9.

Assuming that the angle of an incidence wave from a far-field source is φ, the signal received at the first antenna is

$$s_1(t) = \exp\left\{ j2\pi f_c t \right\} \tag{1.69}$$

and that at the second antenna is

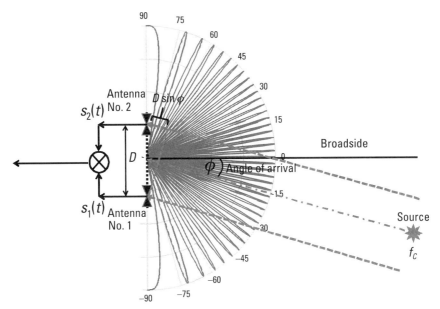

Figure 1.9 A typical interferometric correlation receiver has two separated receiver-channels observing a far-field source.

$$s_2(t) = \exp\left\{ j2\pi f_c(t - \tau) \right\} \qquad (1.70)$$

where f_c is the source frequency and τ is the time delay compared to that at the first antenna determined by

$$\tau = D\sin\varphi/c \qquad (1.71)$$

where φ is the angle of arrival, D is the baseline, and c is the propagation speed.

When the two received signals pass through a complex correlator, that is, a complex multiplier and a lowpass filter, the complex correlator response is

$$C(\varphi) = \left\langle s_1(t) \cdot s_2^*(t) \right\rangle = \exp\left\{ j2\pi f_c \tau \right\} = \exp\left\{ j2\pi f_c D\sin\varphi/c \right\} \quad (1.72)$$

The real part of the correlator output, $\mathrm{Re}[C(\varphi)]$, forms a fringe pattern as shown in Figure 1.9. It indicates that when the angle of the incidence wave φ increases from $-90°$ to $90°$, the correlator response is a sinusoidal type.

The angular velocity of an object is defined by the time derivative of angle

$$\Omega = d\varphi(t)/dt \qquad (1.73)$$

Thus, the angle of the incidence wave can be replaced by $\varphi = \Omega t$. The correlator output response can be rewritten as

$$C(\Omega t) = \exp\left\{ j2\pi D \sin \Omega t / \lambda \right\} \qquad (1.74)$$

When an object passes through the interferometer beam pattern, an oscillation occurs in the correlator response. The oscillation frequency is proportional to the angular velocity of the object.

The instantaneous frequency of the correlator response is the angular velocity induced frequency shift, which is called the interferometric frequency shift:

$$f_{\text{Inf}} = \frac{1}{2\pi} \frac{d}{dt} (2\pi D \sin \Omega t / \lambda) = D\Omega \cos \Omega t / \lambda \qquad (1.75)$$

For the near broadside of the antenna, $\cos \Omega t \approx 1$ and $f_{\text{Inf}} \approx D\Omega/\lambda$. Thus, the interferometric radar response becomes

$$s_{\text{Inf}}(t) = \exp\left\{ j2\pi f_{\text{Inf}} t \right\} \qquad (1.76)$$

which has the same form as the Doppler radar response

$$s_D(t) = \exp\left\{ j2\pi f_D t \right\} \qquad (1.77)$$

where $f_D = 2v_r/\lambda$ is the Doppler frequency shift and v_r is the radial velocity.

The interferometric frequency shift f_{Inf} is proportional to the angular velocity and the Doppler frequency shift f_D is proportional to the radial velocity.

For a broadband source, the effect of bandwidth on the correlator response is the integration over the bandwidth B and given by

$$C(\varphi) = \int_{f_c-B/2}^{f_c+B/2} \exp\left\{ j2\pi f_c D \sin \varphi / c \right\} df$$

$$= \exp\left\{ j2\pi f_c D \sin \varphi / c \right\} \cdot \text{sinc}\left(\pi B D \sin \varphi / c \right) \qquad (1.78)$$

where the sinc function determines the bandwidth-pattern modulated on the fringe pattern [29].

By combining the traditional Doppler measurement of radial velocity with the measurement of angular velocity using the interferometric receiver, the movement of objects can be measured directly, over a wide field of view.

The radial Doppler shift can be measured by a single-channel receiver and the angular Doppler shift uses the dual interferometric receivers. A simulation of a person walking tangential to an interferometric radar for the calculation of the angular velocity induced Doppler shifts is given in [34–36]. The interferometric radar provides a complementary way to Doppler radar in measurement of velocities.

1.11 Research and Applications of Radar Micro-Doppler Signatures

An early study on radar micro-Doppler signature of a walking human was conducted in 1998 managed by the Office of Naval Research [3, 4]. A man was walking toward a radar at normal walking speed. Figure 1.10(a) shows the radar range-Doppler image of the walking person, where the hot spot in the image indicates the body of the person. The articulated motion of arms and legs causes smearing lines across the Doppler direction around the human body. Micro-Doppler signature of the walking person shown in Figure 1.10(b) indicates the Doppler shift of the human body and micro-Doppler shifts of the swinging arms and legs. The body torso's Doppler shift is almost constant with a slightly saw-tooth shape, but the arm' and leg's micro-Doppler shifts are time-varying periodic curves.

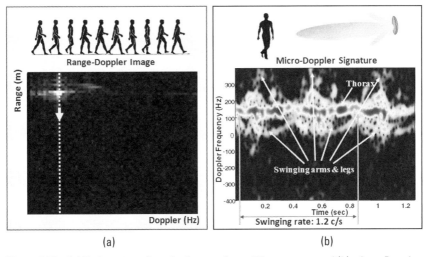

(a)　　　　　　(b)

Figure 1.10　(a) Radar range-Doppler image of a walking person, and (b) micro-Doppler signature of the walking human.

Since then, publications related to micro-Doppler effect in radar have appeared in journals and conference proceedings. Among numerous titles, some of them contribute to theoretical analysis on the micro-Doppler effect, and many others contribute to explore applications of micro-Doppler radar signatures. Also, some micro-Doppler related titles can be found in research dissertations [37–40] and textbooks [41–43].

Motivated by visual surveillance, athletic performance analysis, and biometrics, the extraction and analysis of various human body movements have attracted much attention. Human gait has been studied for a long time in biomedical engineering, sports medicine, physiotherapy, medical diagnosis, and rehabilitation. After early studies on radar micro-Doppler signatures of human gaits, further deeper analysis of radar micro-Doppler signatures for various human motions (such as running, jumping, crawling, and falling) have been conducted [44–53].

With both high-resolution range profiles and high-resolution Doppler spectrogram, the ultrawideband (UWB) radar helps to extract detailed gaiting features, such as balancing legs and swinging arms. The unique characteristics of very detailed features can be used to recognize human activities, such as marching, walking, one-arm swinging, or two-arm swinging. The use of combined micro-Doppler signatures with micro-range features has been reported in [54–58]. The use of ultrafine range resolution and leveraging fine micro-range and micro-Doppler signatures can decompose Doppler and range features based on human body physical component parts.

Research into the use of a UWB radar for human gait research can be found in [58, 59]. The radar provides both high-resolution range profiles and high-resolution Doppler spectrogram to extract detailed micro-Doppler signatures like swinging arms. The unique characteristics of the very detailed signatures can be used to recognize human activities, such as marching, walking, one-arm swinging, or two-arm swinging.

Multistatic radar can observe an object from different aspects and provides multiple aspect-views of its micro-Doppler signature. The combined micro-Doppler signature from multiple radars depends on the topology of the system, the location, and the moving direction of the object. By combining information captured from multiple channels, the location, moving direction, and velocity of the object can be measured. With increased information, the radar performance on target recognition is expected to be improved [60–63].

Micro-Doppler signatures formed from multi-angle observations in a radar network can improve the performance of oblique-angle classification. This issue was explored in [64, 65] by using mutual information to find the degree of importance of features for the target classification.

Because the radar cross-section (RCS) of the human body is small (about 0.5 m^2), radar returns from humans especially at a long distance are very weak. Thus, the radar used for human micro motion study must have adequate transmit power or operate at a short distance. In the real world, radar detection of humans is usually performed in a complex background and clutter environment. Especially, when humans move relatively slow, the intensity of clutter may exceed that from humans. Thus, how to detect weak human signals in clutter becomes an important research topic in human motion analysis.

As we know, higher frequency can have higher Doppler shifts even if the target is in a slow motion. Thus, the use of higher frequency band (such as K-band and W-band) for studying micro-Doppler signatures was conducted. In [66], a W-band at 77-GHz radar was used to observe micro-Doppler signatures of human gait. It was able to recognize multiple persons and identify whether the person is carrying weights.

In studying biological motion perception, it has been found that the motion kinematic of human body parts carries information about human actions, emotions, and even the gender. Therefore, the decomposition of human micro-Doppler signatures became a challenging issue that may lead to the identification of human actions, emotions, and even the gender recognition through micro-Doppler signatures [67, 68].

Ultrasound can also be applied to extract micro-Doppler signatures of moving objects. [69, 70] explored the utility of ultrasonic sensors to distinguish between people and animals walking.

More advanced applications of radar micro-Doppler signatures have been conducted, including recognition of targets in space (such as missile warheads), recognition of the target in the air (such as unmanned aerial vehicles and flying birds), recognition of ground-moving targets (such as vehicles and dismounted combats), recognition of targets underground (such as vital sign detection for trapped human in rubble), and recognition of targets behind a wall.

1.11.1 Micro-Doppler Signatures of Space Targets

Based on the investigation of the signature of a spinning symmetric top, spinning ballistic missile warheads became a natural extended research topic. From micro-Doppler features of missile warheads, some micromotion parameters, such as spin rate, precession rate, nutation angle, and inertia ratio, can be estimated. The precession and nutation of a ballistic missile warhead and the wobble motion of a decoy are two typical micromotions. Their different micro-Doppler signatures may be used to identify a warhead from decoys. It has been found that since the inertial parameters of an object are closely related to the state of its micromotion, the inertial ratio of a rigid cone can serve as an important merit index of the object for the target discrimination [71–81].

1.11.2   Micro-Doppler Signatures of Air Targets

Helicopter identification has become an attractive topic. To identify a helicopter type, besides its shape and size, the number of blades, the length of the blade, and the rotation rate of the rotor are all important features for identifying the helicopter. These parameters can be estimated from micro-Doppler signatures of helicopters.

Micro-Doppler signatures of helicopter rotor blades extracted by monostatic, bistatic, and multistatic radars have been studied [82–85]. They can be used to either estimate rotating parameters of the blades for identification or remove their effect on radar returned signals for imaging.

Other than studying micro-Doppler signatures of conventional helicopter rotor blades, studies of micro-Doppler signatures for multicopters, small unmanned helicopters, and rotary-wing mini unmanned aerial vehicles (UAVs) are also widely conducted. Unlike traditional air targets, small UAVs are smaller in size and fly at slow speeds and low altitudes. These make UAVs easily hidden by complex terrain, difficult to differentiate from birds, and difficult to sense.

Although the RCS of birds is usually small (from −40 dBsm to −10 dBsm), modern surveillance radars can still detect birds from a long distance. On radar screens, high densities of birds can create large numbers of tracks similar to those of mini UAVs or drones. Thus, flying birds may cause false alarms when detecting UAVs. It is mandatory for radars to detect incoming UAVs from flying birds.

There is a need for effective methods to distinguish between birds and mini UAVs. The Doppler spread of bird wing flapping was utilized for identifying flying birds more than 30 years ago [86]. The locomotion of flying birds, especially the elevation and depression of their wings, can generate distinctive micro-Doppler signatures. Therefore, micro-Doppler signatures can be used to discriminate different flying objects and distinguish flying birds from UAVs and other air targets [87–101].

Flapping wings can be clearly seen from the time-varying Doppler spectrum of flying birds. For a Canadian goose, the Doppler shifts from its wings can be 180 Hz higher than those from its body. The bandwidth of the time-varying Doppler spectrum is related to the size of the bird, and this can be used to discriminate birds [93, 94]. Quite different micro-Doppler signatures of two different flapping styles of birds (passerine-like flapping style and swift-like flapping style) were observed. It shows that the passerine-like flapping style has repeated clusters of larger fluctuations [95].

In [96], a K-band CW radar was used to investigate the potential for micro-Doppler signatures in detecting birds. A method that combined the micro-Doppler signatures with range-Doppler images for classifying a single bird and a bird flock was proposed in [97].

In [100, 101], an interferometric radar is discussed in cases where the target is moving tangentially to the radar and the motion relative to the radar is primarily angular. Thus, a measurement of the angular velocity induced micro-Doppler signature is necessary to enable classification regardless of the trajectory of the target.

1.11.3   Micro-Doppler Signatures of Vital Signs

Observing, measuring, and monitoring vital signs, such as heartbeat rate, pulse rate, and breathing rate, are very important not only for health care, but also for finding trapped survivors under rubble or behind barriers. In 1975, J. C. Lin first proposed a microwave technique for measuring respiratory movements of humans and animals [102]. Later, K. M. Chen et al. developed an X-band life-detection radar system for detecting heartbeats and breathing of human subjects [103, 104]. Based on the fact that radar signals returned from trapped survivors can be modulated by physical vibrations due to human breathing and heart beating, radar can be used for finding trapped survivors from a distance without physical contact. Specific vital signs that can be detected by radar include heartbeats, thorax motion by breathing, and even vibration in larynx.

Micro-Doppler signatures of vital signs have been used for noncontact detection of human life [105–110]. CW radar is a simple tool for the noninvasive measuring of respiratory and heartbeat rates from micro-Doppler features [105]. UWB radar can certainly be used to detect heartbeats and observe micro-Doppler signatures of multiple targets at different ranges [106]. In [107], a millimeter-wave radar was used to capture micro-Doppler signatures of human respiration and heartbeat rates at a distance up to 100 meters. For recognizing human breathing and torso bending, the EMD analysis can be used to decompose a multicomponent signal into monocomponent constituents as reported in [108].

1.11.4   Through-the-Wall Radar Micro-Doppler Signatures

The use of radar micro-Doppler signatures to detect and identify targets behind walls has been studied in [111–116]. Through-the-wall radars typically operate at frequencies lower than 5 GHz. Radar signals that travel through walls will experience attenuation and dispersion. Wall losses can reduce the SNR in radar returns and thus limit the maximum radar detectable range. If a maximum detectable range is 50m when a radar operates in an environment without a wall, a 30-dB two-way wall attenuation will decrease the maximum detectable range by a factor of $(10^{-30/10})^{1/4} = 0.18$, or the maximum detectable range becomes 9m. It shows that the maximum detectable range is halved for every additional 12 dB of loss.

Because of the relatively lower frequency used in through-the-wall radars, micro-Doppler frequency shifts can be very low, which makes it difficult to detect objects behind the wall. However, with advanced signal decomposition and processing techniques, radar can still sense human body motion, breathing, and even heartbeats for detecting and monitoring human activities in cases of finding living humans after earthquakes or in explosion scenarios.

The impact of a wall on the micro-Doppler effect has been investigated [111, 112]. It was found that the micro-Doppler effect in the presence of a wall has a similar form as that in free space. The measured aspect angle of an object behind a wall is different to that observed in free space. The measured angle depends on the thickness and dielectric constant of the wall. The change of the instantaneous aspect angle due to the wall will affect radar imaging of objects. However, the presence of a wall does not change the pattern of the micro-Doppler signature of an object. The wall only changes the absolute value of the micro-Doppler signature depending on the wall properties. Therefore, radar micro-Doppler signatures can be used to detect the presence of human beings and their movements behind a wall.

1.11.5   Micro-Doppler Signatures for Indoor Monitoring

Radar is also a prime candidate for monitoring regular and abnormal human activities inside office buildings, homes, schools, and hospitals. Micro-Doppler signatures for indoor monitoring become an increasingly important research topic in home security/safety, home automation, and health status monitoring (activity monitoring, fall detection, and vital sign monitoring).

Indoor human motion includes regular periodic movements (such as walking or running) and nonperiodic movements (such as standing up, sitting down, kneeling, and falling). Such aperiodic events can be an important indication of health, for example, a chronic limp, concussion, dizziness, or even a critical event such as a heart attack. Because such micro-Doppler frequencies are directly associated with the motion and maneuvering behaviors of the human body parts, these micro-Doppler signatures can be used to characterize and classify such motions and maneuvering patterns [117–127]. By carefully analyzing various patterns in the signature, features unique to different activities may be identified and served as a basis for the discrimination and characterization of human movement.

1.11.6 Micro-Doppler Signatures for Hand Gesture Recognition

Hand gestures are hand movements involving fingers, palm, and wrist with the intent of conveying meaningful information and interacting with the environment. Doppler radar sensors have been successfully applied to track hand gestures for human-computer interaction [128–130].

Compared with optical motion capture systems, radar sensors do not require any tag attached to the human hand and easily penetrate plastic or some other materials. Radar sensors are insensitive to the distance, light condition, and background complexity. Thus, they are more robust than optic-based sensors and more suitable for smart gesture recognition and control.

The idea of using Doppler radar for hand gesture recognition was proposed in 2013 [131]. The use of radar micro-Doppler signatures in radar gesture recognition can be found in [132–135]. The micro-Doppler signature of hand gesture is a unique feature of the gesture. Together with the range, velocity, and angle information on fingers, palm, and wrist, the radar sensor has been successfully applied to hand gesture recognition and control.

1.11.7 Micro-Doppler Signatures for Target Classification

Target classification is an extremely important research topic in radar. Commonly used statistical algorithms for target classification include linear discriminant, naïve Bayes classifier, support vector machine (SVM), and kernel machine. Micro-Doppler signatures can be used as features for classification [136–139].

Machine learning is a method from the artificial intelligence that gives "computers the ability to learn without being explicitly programmed" [140]. Deep learning is a subset of machine learning that uses artificial neural networks to learning tasks [141]. A convolutional neural network, such as a deep and feed-forward artificial neural network, has been successfully applied to the target classification. Based on micro-Doppler signatures, deep-learning convolutional neural networks have been successfully applied to classifying human activities [142, 143].

1.11.8   Other Applications of Radar Micro-Doppler Signatures

In the early 2000s, researchers found that large wind turbines can have interference that has a significant effect on radars [144]. Since then, wind turbine clutter has been a new form of radar clutter and interference. Sometimes, wind turbines can cause spurious detection and tracking of aircraft as reported in [145, 146]. The wind turbine features with respect to the wind turbine's position, shape, material, and rotation of blades make radar micro-Doppler signatures complicated [147, 148].

Another interesting application is to use Wi-Fi, FM radio, mobile phone networks, DVB-T (digital video broadcasting terrestrial), or GNSS (global navigation satellite systems) as passive transmitters. The micro-Doppler feature observed in passive radars has been applied to helicopter classification [149–153] and indoor health-care monitoring [154–156].

Micro-Doppler signatures are also used in the radar-based indoor monitoring of human for rehabilitation and assisted living [157, 158]. Because changes in human gait characteristics are associated with the risk of falling [159], it is important to detect human gait abnormalities and monitor alterations in micro-Doppler signatures of human walking [160].

Other interesting application of micro-Doppler signatures is to detect human beings in an urban environment with a nonline-of-sight (NLOS) radar. In urban environments, buildings can generate shadow areas where NLOS radars with multipath propagation can be employed. The presence of the multipaths makes the radar able to detect the NLOS target through diffraction and specular reflection. The radar that may detect targets around street corners is denoted as around-the-corner or behind-the-corner radar [161]. Some experimental results show that micro-Doppler signatures of one or more walking persons can be retrieved from the radar LOS in urban environments [162, 163].

1.12   Organization of the Book

In Chapter 2, the basic mathematics of the micro-Doppler effect in radar, the basic mathematic representations of micromotions, the basic radar scattering from a body with micromotion, the basic mathematics for calculation of micro-Doppler effect, and the bistatic and multistatic micro-Doppler effect are introduced. Then detailed analyses of the micro-Doppler effect in rigid bodies and nonrigid bodies are discussed in Chapters 3 and 4.

Because of the recent progress on the radar sensing of human vital signs and hand gestures, applications of radar micro-Doppler signatures to vital sign detection and to human hand gesture recognition are introduced in Chapters 5 and 6, respectively.

Based on currently developed highly integrated low-power and compact radar, the CW and FMCW radars, which offer a great deal on the system compactness, flexibility, and low costs, can be used as a micro-Doppler radar for sensing of micromotions. An overview of the requirement and system architecture is given in Chapter 7.

From the biological motion perception and the relationship between biological motion information and micro-Doppler signatures, Chapter 8 discusses how to analyze and interpret micro-Doppler signatures.

Chapter 9 summarizes the micro-Doppler effect in radar introduced in this book and lists some challenges and perspectives in micro-Doppler research.

References

[1]   Eden, A., *The Search for Christian Doppler*, New York: Springer-Verlag, 1992.

[2]   Gill, T. P., *The Doppler Effect: An Introduction to the Theory of the Effect*, London: Logos Press/Academic Press, 1965.

[3]   Chen, V. C., "Analysis of Radar Micro-Doppler Signature with Time-Frequency Transform," *Proc. of the IEEE Workshop on Statistical Signal and Array Processing* (SSAP), Pocono, PA, 2000, pp. 463–466.

[4]   Chen, V. C. and H. Ling, *Time-Frequency Transforms for Radar Imaging and Signal Analysis*, Norwood, MA: Artech House, 2002.

[5]   Chen, V. C., et al., "Micro-Doppler Effect in Radar: Phenomenon, Model, and Simulation Study," *IEEE Transactions on Aerospace and Electronics Systems*, Vol. 42, No. 1, 2006, pp. 2–21.

[6]   Van Bladel, J., *Relativity and Engineering*, New York: Springer, 1984.

[7]   Willis, N. J., *Bistatic Radar*, 2nd ed., Raleigh, NC: SciTech Publishing, 2005.

[8]   Chernyak, V., *Fundamentals of Multisite Radar Systems*, Amsterdam: Gordon and Breach Science Publishers, 1998.

[9]   Rife, D. C., and R. R. Boorstyn, "Single Tone Parameter Estimation from Discrete-Time Observations," *IEEE Transactions on Information Theory*, Vol. 20, No. 5, 1974, pp. 591–598.

[10]   Kay, S. M., and S. L. Marple, "Spectrum Analysis: A Modern Perspective," *Proceedings of IEEE*, Vol. 69, No. 11, 1981, pp. 1380–1419.

[11]   Marple, S. L., *Digital Spectral Analysis with Applications*, Englewood Cliffs, NJ: Prentice-Hall, 1987.

[12]   Kay, S. M., "A Fast and Accurate Single Frequency Estimator," *IEEE Transactions on Acoustics Speech, Signal Processing*, Vol. 37, No. 12, 1989, pp. 1987–1990.

[13]   Rao, C. R., "Information and Accuracy Attainable in the Estimation of Statistical Parameters," *Bulletin of the Calcutta Mathematical Society*, Vol. 37, 1945, pp. 81–91.

[14]   Kay, S. M., *Fundamentals of Statistical Signal Processing: Estimation Theory*, Upper Saddle River, NJ: Prentice Hall, 1993.

[15]   Gupta, M. S., "Definition of Instantaneous Frequency and Frequency Measurability," *American Journal of Physics*, Vol. 43, No. 12, 1975, pp. 1087–1088.

[16]   Huang, N. E., et al., "The Empirical Mode Decomposition and the Hilbert Spectrum for Nonlinear and Non-Stationary Time Series Analysis," *Proc. Roy. Soc. London*, Ser. A, Vol. 454, 1998, pp. 903–995.

[17]   Boashash, B., "Estimating and Interpreting the Instantaneous Frequency of a Signal—Part 2: Algorithms and Applications," *Proceedings of IEEE*, Vol. 80, No. 4, 1992, pp. 540–568.

[18]   Olhede, S., and A. T. Walden, "The Hilbert Spectrum Via Wavelet Projections," *Proc. R. Soc. London*, Ser. A, Vol. 460, 2004, pp. 955–975.

[19]   Rilling, G., and P. Flandrin, "Bivariate Empirical Mode Decomposition," *IEEE Signal Processing Letter*, Vol. 14, No. 12, 2007, pp. 936–939.

[20]   Tanaka, T., and D. P. Mandic, "Complex Empirical Mode Decomposition," *IEEE Signal Processing Letter*, Vol. 14, No. 2, 2007, pp. 101–104.

[21]   CEMD: http://perso.ens-lyon.fr/patrick.flandrin.

[22]   Gabor, D., "Theory of Communication," *Journal of IEE (London)*, Vol. 93, Part III, No. 26, 1946, pp. 429–457.

[23]   Cohen, L., *Time-Frequency Analysis*, Prentice Hall, Englewood Cliffs, NJ, 1995.

[24]   Qian, S., and D. Chen, *Introduction to Joint Time-Frequency Analysis: Methods and Applications*, Prentice Hall, Englewood Cliffs, NJ: Prentice Hall, 1996.

[25]   Time-Frequency Toolbox: http://tftb.nongnu.org/.

[26]   Shirman, Y. D., *Computer Simulation of Aerial Target Radar Scattering, Recognition, Detection, and Tracking*, Norwood, MA: Artech House, 2002.

[27]   Gorshkov, S. A., et al., *Radar Target Backscattering Simulation: Software and User's Manual*, Norwood, MA: Artech House, 2002.

[28]   Chen, V. C., C. -T. Lin, and W. P. Pala, "Time-Varying Doppler Analysis of Electromagnetic Backscattering from Rotating Object," *The IEEE Radar Conference Record*, Verona, NY, April 24–27, 2006, pp. 807–812.

[29]   Nanzer, J. A., "Millimeter-Wave Interferometric Angular Velocity Detection," *IEEE Transactions on Microwave Theory and Techniques*, Vol. 58, No. 12, 2010, pp. 4128–4136.

[30]   Nanzer, J. A., "On the Resolution of the Interferometric Measurement of the Angular Velocity of Moving Objects," *IEEE Transactions on Aerospace and Electronics Systems*, Vol. 60, No. 11, 2012, pp. 5356–5363.

[31]   Nanzer, J. A., and K. Zilevu, "Dual Interferometric-Doppler Measurements of the Radial and Angular Velocity of Humans," *IEEE Transactions on Antennas and Propagation*, Vol. 62, No. 3, 2014, pp. 1513–1517.

[32]   Kraus, J. D., *Radio Astronomy*, New York: McGraw-Hill, 1966.

[33]   Thompson, A. R., J. M. Moran, and G. W. Swenson, Jr., *Interferometry and Synthesis in Radio Astronomy*. New York: Wiley, 2001.

[34]   Nanzer, J. A., "Interferometric Measurement of the Angular Velocity of Moving Human," *Proc. SPIE 8361 Radar Sensor Technology XVI*, 2012, 836102.

[35]   Nanzer, J. A., "Simulations of the Millimeter-Wave Interferometric Signature of Walking Humans," *IEEE Antennas and Propagation Society International Symposium (APSURSI)*, 2012.

[36]   Nanzer, J. A., "Micro-Motion Signatures in Radar Angular Velocity Measurement," *IEEE Radar Conference*, Philadelphia, PA, May 2016.

[37]   Anderson, M. G., "Design of Multiple Frequency Continuous Wave Radar Hardware and Micro Doppler Based Detection and Classification Algorithms," Ph.D. Dissertation, University of Texas at Austin, 2008.

[38]   Smith, G. E., "Radar Target Micro-Doppler Signature Classification," Ph.D. Dissertation, University College London, 2008.

[39]   Ghaleb, A., "Micro-Doppler Analysis of Time Varying Targets in Radar Imaging," Ph.D. Dissertation, University Télécom ParisTech, February 2009.

[40]   Molchanov, P., "Radar Target Classification by Micro-Doppler Contributions," Ph.D. Dissertation, Tampere University of Technology, Finland, 2014.

[41]   Chen, V. C., D. Tahmoush, and W. J. Miceli, (eds.), *Radar Micro-Doppler Signature: Processing and Applications*, IET, London, UK, 2014.

[42]   Zhang, Q., Y. Luo, and Y. A. Chen, *Micro-Doppler Characteristics of Radar Targets*, New York: Elsevier, 2017.

[43]   Amin, M. G., (ed.), *Radar for Indoor Monitoring Detection, Classification, and Assessment*, Boca Raton, FL: CRC Press, 2018.

[44]   Chen, V. C., et al., "Analysis of Micro-Doppler Signatures," *IEE Proceedings: Radar, Sonar and Navigation*, Vol. 150, No. 4, 2003, pp. 271–276.

[45]   Chen, V. C., "Advances in Applications of Radar Micro-Doppler Signatures," *2014 IEEE Conference on Antenna Measurements & Applications (CAMA)*, 2014, pp. 1–4.

[46] Thayaparan, T., et al., "Analysis of Radar Micro-Doppler Signatures from Experimental Helicopter and Human Data," *IET Radar, Sonar and Navigation*, Vol. 1, No. 4, 2007, pp. 289–299.

[47] Anderson, M. G., and R. L. Rogers, "Micro-Doppler Analysis of Multiple Frequency Continuous Wave Radar Signatures," *Proceedings of SPIE: Radar Sensor Technology XI*, Vol. 6547, 2007, 65470A.

[48] Ram, S. S., et al., "Doppler-Based Detection and Tracking of Humans in Indoor Environments," *Journal of the Franklin Institute*, Vol. 345, No. 6, 2008, pp. 679–699.

[49] Thayaparan, T., et al., "Micro-Doppler-Based Target Detection and Feature Extraction in Indoor and Outdoor Environments," *Journal of the Franklin Institute*, Vol. 345, No. 6, 2008, pp. 700–722.

[50] van Dorp, P., "Identifying Human Movements Using Micro-Doppler Features," Chapter 6 in *Radar Micro-Doppler Signature: Processing and Applications*, V. C. Chen, D. Tahmoush, and W. J. Miceli (eds.), Radar Series 34, IET, 2014, pp. 139–185.

[51] Wang, Y. -Z., Q. -H. Liu, and A. E. Fathy, "CW and Pulse-Doppler Radar Processing Based on FPGA for Human Sensing Applications," *IEEE Transactions on Geoscience and Remote Sensing*, Vol. 51, No. 5, 2013, pp. 3097–3107.

[52] Ghaleb, A., and L. Vignaud, "Range and Micro-Doppler Analysis of Human Motion Using High Resolution Experimental HYCAM Radar," Chapter 4 in *Radar Micro-Doppler Signature: Processing and Applications*, V. C. Chen, D. Tahmoush, and W. J. Miceli (eds.), Radar Series 34, IET, 2014, pp. 69–96.

[53] Narayanan, R. M., and M. Zenaldin, "Radar Micro-Doppler Signatures of Various Human Activities," *IET Radar, Sonar & Navigation*, Vol 9, No. 9, October 2015, pp. 1205–1215.

[54] Fogle, O. R., and B. D. Rigling, "Micro-Range/Micro-Doppler Decomposition of Human Radar Signatures," *IEEE Transactions on Aerospace and Electronics Systems*, Vol. 48, No. 4, 2012, pp. 3058–3072.

[55] Fogle, O. R., and B. D. Rigling, "Analysis of Human Signatures Using High-Range Resolution Micro-Doppler Radar," Chapter 3 in *Radar Micro-Doppler Signature: Processing and Applications*, V. C. Chen, D. Tahmoush, and W. J. Miceli, (eds.), Radar Series 34, IET, 2014, pp. 97–137.

[56] Cammenga, Z. A., G. E. Smith, and C. J. Baker, "Combined High Range Resolution and Micro-Doppler Analysis of Human Gait," *Proceedings of 2015 IEEE Radar Conference*, 2015, pp. 1038–1043.

[57] Ghaleb, A., L. Vignaud, and J. M. Nicolas, "Micro-Doppler Analysis of Wheels and Pedestrians in ISAR Imaging," *IET Signal Processing*, Vol. 2, No. 3, 2008, pp. 301–311.

[58] Vignaud, L., et al., "Radar High Resolution Range & Micro-Doppler Analysis of Human Motions," *Proceedings of IEEE 2009 International Radar Conference*, 2009, pp. 1–6.

[59] Wang, Y., and A. E. Fathy, "Micro-Doppler Signatures for Intelligent Human Gait Recognition Using a UWB Impulse Radar," *IEEE 2011 International Symposium on Antennas and Propagation (APSURSI)*, 2011, pp. 2103–2106.

[60] Smith, G. E., K. Woodbridge, and C. Baker, "Multistatic Micro-Doppler Signature of Personnel," *IEEE Radar Conference*, Rome, Italy, May 2008.

[61] Chen, V. C., A. des Rosiers, and R. Lipps, "Bi-Static ISAR Range-Doppler Imaging and Resolution Analysis," *IEEE Radar Conference*, Pasadena, CA, 2009.

[62] Smith, G. E., et al., "Multistatic Micro-Doppler Radar Signatures of Personnel Targets," *IET Signal Processing*, Vol. 4, No. 3, 2010, pp. 224–233.

[63] Smith, G., and C. Baker, "Multistatic Micro-Doppler Signature Processing," Chapter 9 in *Radar Micro-Doppler Signature: Processing and Applications*, V. C. Chen, D. Tahmoush, and W. J. Miceli, (eds.), Radar Series 34, IET, 2014, pp. 241–272.

[64] Tekeli, B., et al., "Classification of Human Micro-Doppler in Radar Network," *Proceedings of 2013 IEEE Radar Conference*, Ottawa, Canada, 2013.

[65] Tekeli, B., S. Z. Gurbuz, and M. Yuksel, "Mutual Information of Features Extracted from Human Micro-Doppler," *2013 21st Signal Processing and Communications Applications Conference*, 2013, pp. 1–4.

[66] Bjorklund, S., et al., "Millimeter-Wave Radar Micro-Doppler Signatures of Human Motion," *Proceedings of 2011 International Radar Symposium* (*IRS*), 2011, pp. 167–174.

[67] Garreau, G., et al., "Gait-Based Person and Gender Recognition Using Micro-Doppler Signatures," *IEEE Biomedical Circuits and Systems Conference* (*BioCAS*), 2011, pp. 444–447.

[68] Ahmed, M. H., and A. T. Sabir, "Human Gender Classification Based on Gait Features Using Kinect Sensor," *2017 3rd IEEE International Conference on Cybernetics*, 2017, pp. 1–5.

[69] Damarla, T., et al., "Classification of Animals and People Ultrasonic Signatures," *IEEE Sensor Journal*, Vol. 13, No. 5, 2013, pp. 1464–1472.

[70] Murray, T. S., et al., "Bio-Inspired Human Action Recognition with a Micro-Doppler Sonar System," *IEEE Access*, Vol. 6, 2017, pp. 28388–28403.

[71] Ning, C., et al., "Modeling and Simulation of Micro-Motion in the Complex Warhead Target," *Proceedings of SPIE—Second International Conference on Space Information Technology*, Vol. 6795, 2007.

[72] Sun, H. X., and Z. Liu, "Micro-Doppler Feature Extraction for Ballistic Missile Warhead," *Proc. of the 2008 IEEE International Conference on Information and Automation* (*ICIA*), 2008, pp. 1333–1336.

[73] Guo, K.Y. and X. Q. Sheng, "A Precise Recognition Approach of Ballistic Missile Warhead and Decoy," *Journal of Electromagnetic Waves and Applications*, Vol. 23, No. 14-15, 2009, pp. 1867–1875.

[74]   Gao, H., et al., "Micro-Doppler Signature Extraction from Ballistic Target with Micro-Motions," *IEEE Transactions on Aerospace and Electronics Systems*, Vol. 46, No. 4, 2010, pp. 1968–1982.

[75]   Wang, T., et al., "Estimation of Precession Parameters and Generation of ISAR Images of Ballistic Missile Targets," *IEEE Transactions on Aerospace and Electronics Systems*, Vol. 46, No. 4, 2010, pp. 1983–1995.

[76]   Lei, P., J. Wang, and J. Sun, "Analysis of Radar Micro-Doppler Signatures from Rigid Targets in Space Based on Inertial Parameters," *IET Radar, Sonar, Navigation*, Vol. 5, No. 2, 2011, pp. 93–102.

[77]   Li, M., and Y. S. Jiang, "Feature Extraction of Micro-Motion Frequency and the Maximum Wobble Angle in a Small Range of Missile Warhead Based on Micro-Doppler Effect," *Optics and Spectroscopy*, Vol. 117, No. 5, 2014, pp. 832–838.

[78]   Choi, I. O., "Estimation of the Micro-Motion Parameters of a Missile Warhead Using a Micro-Doppler Profile," *2016 IEEE Radar Conference*, 2016, pp. 1–5.

[79]   Shi, Y. C., et al., "A Coning Micro-Doppler Signals Separation Algorithm Based on Time-Frequency Information," *2017 IEEE International Conference on Signal Processing, Communications and Computing (ICSPCC)*, 2017, pp. 1–5.

[80]   Persico, A. R., et al., "On Model, Algorithms, and Experiment for Micro-Doppler-Based Recognition of Ballistic Targets," *IEEE Transactions on Aerospace and Electronic Systems*, Vol. 53, No. 3, 2017, pp. 1088–1108.

[81]   Zhou, Y., "Micro-Doppler Curves Extraction and Parameters Estimation for Cone-Shaped Target with Occlusion Effect," *IEEE Sensors Journal*, Vol. 18, No. 7, 2018, pp. 2892–2902.

[82]   Johnsen, T., K. E. Olsen, and R. Gundersen, "Hovering Helicopter Measured by Bi-/Multistatic CW Radar," *Proceedings of IEEE 2003 Radar Conference*, 2003, pp. 165–170.

[83]   Olsen, K. E., et al., "Multistatic and/or Quasi Monostatic Radar Measurements of Propeller Aircrafts," *Proceedings of 2007 IET International Radar Conference*, 2007, pp. 1–6.

[84]   Cilliers, A., and W. Nel, "Helicopter Parameter Extraction Using Joint Time-Frequency and Tomographic Techniques," *IEEE 2008 International Conference on Radar*, Adelaide, Australia, 2008, pp. 2–5.

[85]   Molchanov, P., et al., "Aerial Target Classification by Micro-Doppler Signatures and Bicoherence-Based Features," *Proceedings of the 9th European Radar Conference*, Amsterdam, The Netherlands, 2012, pp. 214–217.

[86]   Vaughn, C. R., "Birds and Insects as Radar Targets: A Review," *Proceedings of the IEEE*, Vol. 73, No. 2, 1985, pp. 205–227.

[87]   Singh, A. K., and Y. -H. Kim, "Automatic Measurement of Blade Length and Rotation Rate of Drone Using W-Band Micro-Doppler Radar," *IEEE Sensors Journal*, Vol. 18, No. 5, 2018, pp. 1895–1902.

[88]   Rahman, S., and D. Robertson, "Time-Frequency Analysis of Millimeter-Wave Radar Micro-Doppler Data from Small UAVs," *2017 Sensor Signal Processing for Defense Conference*, 2017, pp. 1–5.

[89]   Fuhrmann, L., et al., "Micro-Doppler Analysis and Classification of UAVs at Ka Band," *2017 18th International Radar Symposium* (*IRS*), June 2017.

[90]   Alabaster, C. M., and E. J. Hughes, "Is It a Bird or Is It a Plane?" *Proceedings of IEEE 6th International Conference on Waveform Diversity & Design*, Lihue, HI, 2012, pp. 007–012.

[91]   Molchanov, P., et al., "Classification of Small UAVs and Birds by Micro-Doppler Signatures," *International Journal of Microwave and Wireless Technologies*, Vol. 6, No. 3-4, 2014, pp. 435–444.

[92]   Ritchie, M., et al., "Monostatic and Bistatic Radar Measurements of Birds and Micro-Drone," *2016 IEEE Radar Conference*, Philadelphia, PA, 2016, pp. 1–5.

[93]   Spruyt, J. A., and P. van Dorp, *Detection of Birds by Radar*, TNO Report, FEL-95-A244, TNO Physics and Electronics Laboratory, 1996.

[94]   Flock, W. L., and J. L. Green, "The Detection and Identification of Birds in Flight Using Coherent and Noncoherent Radars," *Proceedings of IEEE*, Vol. 62, 1974, pp. 745–753.

[95]   Zaugg, S., et al., "Automatic Identification of Bird Targets with Radar Via Patterns Produced by Wing Flapping," *Journal of the Royal Society Interface*, Vol. 5, No. 26, 2008, pp. 1041–1053.

[96]   Torvik, B., K. E. Olsen, and H. Griffiths, "K-Band Radar Signature Analysis of a Flying Mallard Duck," *2013 14th International Radar Symposium*, Dresden, Germany, 2013, Vol. 2, pp. 584–591.

[97]   Ozcan, A. H., et al., "Micro-Doppler Effect Analysis of Single Bird and Bird Flock for Linear FMCW Radar," *2012 20th Signal Processing and Communications Application Conference*, 2012.

[98]   Hoffmann, F., et al., "Micro-Doppler Based Detection and Tracking of UAVs with Multistatic Radar," *Proceedings of 2016 IEEE Radar Conference*, 2016, pp. 1–6.

[99]   Kim, B. K., H. -S. Kang and S. -O. Park, "Experimental Analysis of Small Drone Polarimetry Based on Micro-Doppler Signature," *IEEE Geoscience and Remote Sensing Letters*, Vol. 14, No. 10, 2017, pp. 1670–1674.

[100] Jian, M., Z. Z. Lu, and V. C. Chen, "Experimental Study on Radar Micro-Doppler Signatures of Unmanned Aerial Vehicles," *Proceedings of 2017 IEEE Radar Conference*, 2017, pp. 854–857.

[101] Nanzer, J. A., and V. C. Chen, "Microwave Interferometric and Doppler Radar Measurements of a UAV," *Proceedings of 2017 IEEE Radar Conference*, 2017, pp. 1628–1633.

[102] Lin, J. C., "Noninvasive Microwave Measurement of Respiration," *Proceedings of the IEEE*, Vol. 63, No.10, 1975, pp. 1530–1530.

[103] Chen, K. M., et al., "An X-Band Microwave Life-Detection System," *IEEE Transactions on Biomedical Engineering*, Vol. 33, No. 7, 1986, pp. 697–702.

[104] Chen, K. M., et al., "Microwave Life-Detection Systems for Searching Human Subjects Under Earthquake Rubble or Behind Barrier," *IEEE Transactions on Biomedical Engineering*, Vol. 27, No. 1, 2000, pp. 105–114.

[105] Salmi, J., O. Luukkonen, and V. Koivunen, "Continuous Wave Radar Based Vital Sign Estimation: Modeling and Experiments," *Proceedings of IEEE 2012 Radar Conference*, 2012, pp. 564–569.

[106] Shirodkar, S., et al., "Heart-Beat Detection and Ranging Through a Wall Using Ultrawide Band Radar," *International Conference on Communications and Signal Processing*, 2011, pp. 579–583.

[107] Moulton, M. C., et al., "Micro-Doppler Radar Signatures of Human Activity," *Proceedings of SPIE*, Vol. 7837, 2010, pp. 78370L-1–78370L-7.

[108] Narayanan, R. M., "Earthquake Survivor Detection Using Life Signals from Radar Micro-Doppler," *Proceedings of the 1st International Conference on Wireless Technologies for Humanitarian Relief*, 2011, pp. 259–264.

[109] Gu, C. -Z, et al., "Instrument-Based Noncontact Doppler Radar Vital Sign Detection System Using Heterodyne Digital Quadrature Demodulation Architecture," *IEEE Transactions on Instrumentation and Measurement*, Vol. 59, No. 6, 2010, pp. 1580–1588.

[110] Li, J., et al., "Advanced Signal Processing for Vital Sign Extraction with Applications in UWB Radar Detection of Trapped Victims in Complex Environments," *IEEE Journal of Selected Topics in Applied Earth Observation and Remote Sensing*, Vol. 7, No. 3, 2014, pp. 783–791.

[111] Fairchild, D. P., et al., "Through-the-Wall Micro-Doppler Signatures," Chapter 5 in *Radar Micro-Doppler Signature: Processing and Applications*, V. C. Chen, D. Tahmoush, and W. J. Miceli (eds.), Radar Series 34, IET, 2014, pp. 97–137.

[112] Ram, S. S., et al., "Simulation and Analysis of Human Micro-Dopplers in Through-Wall Environments," *IEEE Transactions on Geoscience and Remote Sensing*, Vol. 48, No. 4, 2010, pp. 2015–2023.

[113] Chen, V. C., et al., "Radar Micro-Doppler Signatures for Characterization of Human Motion," Chapter 15 in *Through-The-Wall Radar Imaging*, M. G. Amin, (ed.), Boca Raton, FL: CRC Press/Taylor & Francis Group, 2011.

[114] Liu, X., H. Leung, and G.A. Lampropoulos, "Effects of Non-Uniform Motion in Through-the-Wall SAR Imaging," *IEEE Transactions on Antenna and Propagation*, Vol. 57, No. 11, 2009, pp. 3539–3548.

[115] Lubecke, V. M., et al., "Through-The-Wall Radar Life Detection and Monitoring," *Proceedings of the 2007 IEEE Microwave Theory and Techniques Symposium*, Honolulu, HI, 2007, pp. 769–772.

[116] Bugaev, A. S., et al., "Through Wall Sensing of Human Breathing and Heart Beating by Monochromatic Radar," *Proceedings of the 10th International Conference on Ground Penetrating Radar*, Delft, the Netherlands, 2004, pp. 291–294.

[117] Ram, S. S., S. Z. Gurbuz, and V. C. Chen, "Modeling and Simulation of Human Motion for Micro-Doppler Signatures," Chapter 3 in *Radar for Indoor Monitoring: Detection, Classification, and Assessment*, M. G. Amin, (ed.), Boca Raton, FL: CRC Press/Taylor & Francis Group, 2018, pp. 39–69.

[118] Zhang, Y. M., and D. K. C. Ho, "Continuous-Wave Doppler Radar for Fall Detection," Chapter 4 in *Radar for Indoor Monitoring: Detection, Classification, and Assessment*, M. G. Amin, (ed.), Boca Raton, FL: CRC Press/Taylor & Francis Group, 2018, pp. 71–93.

[119] Griffiths, H., M. Ritchie, and F. Fioranelli, "Bistatic Radar Configuration for Human Body and Limb Motion Detection and Classification," Chapter 8 in *Radar for Indoor Monitoring: Detection, Classification, and Assessment*, M. G. Amin, (ed.), Boca Raton, FL: CRC Press/Taylor & Francis Group, 2018, pp. 179–198.

[120] Seifert, A. -K., M. G. Amin, and A. M. Zoubir, "Radar Monitoring of Humans with Assistive Walking Devices," Chapter 12 in *Radar for Indoor Monitoring: Detection, Classification, and Assessment*, M. G. Amin, (ed.), Boca Raton, FL: CRC Press/Taylor & Francis Group, 2018, pp. 271–300.

[121] Wu, Q., et al., "Radar-Based Fall Detection Based on Doppler Time-Frequency Signatures for Assisted Living," *IET Radar, Sonar and Navigation*, Vol. 9, No. 2, 2015, pp. 164–172.

[122] Li, C., et al., "A Review on Recent Advances in Doppler Radar Sensors for Noncontact Healthcare Monitoring," *IEEE Transactions on Microwave Theory and Techniques*, Vol. 61, No. 5, 2013, pp. 2046–2060.

[123] Thayaparan, T., L. Stankovic, and I. Djurovic, "Micro-Doppler-Based Target Detection and Feature Extraction in Indoor and Outdoor Environments," *Journal of the Franklin Institute*, Vol. 345, No. 6, 2008, pp. 700–722.

[124] Fioranelli, F., M. Ritchie, and H. Griffiths, "Bistatic Human Micro-Doppler Signatures for Classification of Indoor Activities," *Proceedings of IEEE 2017 Radar Conference*, 2017, pp. 610–615.

[125] Gurbuz, S. Z., et al., "Micro-Doppler-Based In-Home Aided and Unaided Walking Recognition with Multiple Radar and Sonar Systems," *IET Radar, Sonar and Navigation*, Vol. 11, No. 1, 2017, pp. 107–115.

[126] Chen, Q. C., et al., "Joint Fall and Aspect Angle Recognition Using Fine-Grained Micro-Doppler Classification," *Proceedings of IEEE 2017 Radar Conference*, 2017, pp. 912–916.

[127] Vishwakarma, S., and S. S. Ram, "Dictionary Learning For Classification of Indoor Micro-Doppler Signatures Across Multiple Carriers," *Proceedings of IEEE 2017 Radar Conference*, 2017, pp. 992–997.

[128] Molchanov, P., et al., "Short-Range FMCW Monopulse Radar for Hand-Gesture Sensing," *Proceedings of the 2015 IEEE Radar Conference*, Arlington, VA, May 10–15, 2015.

[129] Wang, S., et al., "Interacting with Soli: Exploring Fine-Grained Dynamic Gesture Recognition in the Radio-Frequency Spectrum," *Proceedings of the 29th Annual Symposium on User Interface Software and Technology* (UIST), October 16–19, 2016, Tokyo, Japan, 2016, pp. 851–860.

[130] Lien, J., et al., "Soli: Ubiquitous Gesture Sensing with Millimeter Wave Radar," *ACM Transactions on Graphics*, Vol. 35, No. 4, Article 142, 2016, pp. 1–19.

[131] Zheng, C., et al., "Doppler Bio-Signal Detection Based Time-Domain Hand Gesture Recognition," *Proceedings of the 2013 IEEE MTT-S International Microwave Workshop Series on RF and Wireless Technologies for Biomedical and Healthcare Applications* (IMWS-BIO); Singapore, December 9–11, 2013.

[132] Wan, Q., et al., "Gesture Recognition for Smart Home Applications Using Portable Radar Sensors," *Proceedings of the 2014 36th Annual International Conference of the IEEE Engineering in Medicine and Biology Society* (EMBC); Chicago, IL, August 26–30, 2014.

[133] Molchanov, P., et al., "Multi-Sensor System for Driver's Hand-Gesture Recognition," *Proceedings of the 2015 11th IEEE International Conference and Workshops on Automatic Face and Gesture Recognition*, Ljubljana, Slovenia, May 4–8, 2015.

[134] Kim, Y., and B. Toomajian, "Hand Gesture Recognition Using Micro-Doppler Signatures with Convolutional Neural Network," *IEEE Access*, Vol. 4, 2016, pp. 7125–7130.

[135] Li, G., et al., "Sparsity-Driven Micro-Doppler Feature Extraction for Dynamic Hand Gesture Recognition," *IEEE Transactions on Aerospace and Electronic Systems*, 2017.

[136] Zabalza, J., et al., "Robust PCA Micro-Doppler Classification Using SVM on Embedded Systems," *IEEE Transactions on Aerospace and Electronic Systems*, Vol. 50, No. 3, 2015, pp. 2304–2310.

[137] De Wit, J. J. M., R. Harmanny, and P. Molchanov, "Radar Micro-Doppler Feature Extraction Using the Singular Value Decomposition," *Proceedings of 2014 International Radar Conference*, Lille, France, 2014, pp. 1–6.

[138] Vishwakarma, S., and S. S. Ram, "Dictionary Learning for Classification of Indoor Micro-Doppler Signatures Across Multiple Carriers," *Proceedings of 2017 IEEE Radar Conference*, 2017, pp. 992–997.

[139] Bjorklund, S., H. Petersson, and G. Hendeby, "Features for Micro-Doppler Based Activity Classification," *IET Radar, Sonar and Navigation*, Vol. 9, No. 9, 2015, pp. 1181–1187.

[140] Samuel, A., "Some Studies in Machine Learning Using the Game of Checkers," *IBM Journal of Research and Development*, Vol. 3, No. 3, 1959, pp. 210–229.

[141] LeCun, Y., and Y. Bengio, "Convolutional Networks for Images, Speech, and Time-Series," in M. A. Arbib, (ed.), *The Handbook of Brain Theory and Neural Networks*. Cambridge, MA: MIT Press, 1995.

[142] Kim, Y., and T. Moon, "Human Detection and Activity Classification Based on Micro-Doppler Signatures Using Deep Convolutional Neural Networks," *IEEE Geoscience and Remote Sensing Letters*, Vol. 13, No. 1, 2016, pp. 8–12.

[143] Molchanov, P., et al., "Hand Gesture Recognition with 3D Convolutional Neural Networks," *2015 IEEE Computer Society Conference on Computer Vision and Pattern Recognition Workshop*, 2015, pp. 1–7.

[144] Poupart, G. J., "Wind Farms Impact on Radar Aviation Interests," QinetiQ Ltd, Tech. Rep. W/14/00614/00/REP, 2003.

[145] Buterbaugh, A., et al., "Dynamic Radar Cross Section and Radar Doppler Measurements of Commercial General Electric Windmill Power Turbines Part 2: Predicted and Measured Doppler Signatures," *AMTA Symposium*, St. Louis, MO, 2007.

[146] Kent, B. M., et al., "Dynamic Radar Cross Section and Radar Doppler Measurements of Commercial General Electric Windmill Power Turbines Part 1: Predicted and Measured Radar Signatures," *IEEE Antennas and Propagation Magazine*, April 2008, pp. 211–219.

[147] Kong, F., Y. Zhang, and R. Palner, "Radar Micro-Doppler Signature of Wind Turbines," Chapter 12 in *Radar Micro-Doppler Signature: Processing and Applications*, V. C. Chen, D. Tahmoush, and W. J. Miceli, (eds.), Radar Series 34, IET, 2014, pp. 345–381.

[148] Krasnov, O. A., and A. G. Yarovoy, "Radar Micro-Doppler of Wind-Turbines: Simulation and Analysis Using Slowly Rotating Linear Wired Constructions," *2014 11th European Radar Conference*, 2014, pp. 73–76.

[149] Clemente, C., and J. J. Soraghan, "Passive Bistatic Radar for Helicopters Classification: A Feasibility Study," *Proceedings of 2012 IEEE Radar Conference*, 2012, pp. 946–949.

[150] Baczyk, M. K., et al., "Micro-Doppler Signatures of Helicopters in Multistatic Passive Radars," *IET Radar, Sonar, and Navigation*, Vol. 9, No. 9, 2015, pp. 1276–1283.

[151] Clemente, C., et al., "GNSS Based Passive Bistatic Radar for Micro-Doppler Based Classification of Helicopters: Experimental Validation," *Proceedings of 2015 IEEE Radar Conference*, 2015, pp. 1104–1108.

[152] Clemente, C., and J. J. Soraghan, "GNSS-Based Passive Bistatic Radar for Micro-Doppler Analysis of Helicopter Rotor Blades," *IEEE Transactions on Aerospace and Electronics Systems*, Vol. 50, No. 1, 2014, pp. 491–500.

[153] Xia, P., et al., "Investigations Toward Micro-Doppler Effect in Digital Broadcasting Based Passive Radar," *IET 2015 International Radar Conference*, 2015, pp. 1–5.

[154] Li, W., B. Tan, and R. Piechocki, "Passive Radar for Opportunistic Monitoring in E-Health Applications," *IEEE Journal of Translational Engineering in Health and Medicine*, Vol. 6, 2018, Article No. 2800210.

[155]   Chen, Q.-C., et al., "Activity Recognition Based on Micro-Doppler Signature with In-Home WiFi," *2016 IEEE 18th International Conference on e-Health Networking, Applications and Services*, 2016, pp. 1–6.

[156]   Li, W., B. Tan, and R. Piechocki, "Non-Contact Breathing Detection Using Passive Radar," *2016 IEEE International Conference on Communications*, 2016, pp. 1–6.

[157]   Seifert, A. -K., M. G. Amin, and A. M. Zoubir, "New Analysis of Radar Micro-Doppler Gait Signatures for Rehabilitation and Assisted Living," *2017 IEEE International Conference on Acoustics, Speech and Signal Processing* (ICASSP), 2017, pp. 4004–4008.

[158]   Seifert, A. -K., A. M. Zoubir, and M. G. Amin, "Radar-Based Human Gait Recognition in Cane-Assisted Walks," *Proceedings of 2017 IEEE Radar Conference*, 2017, Seattle, WA, pp. 1428–1433.

[159]   Barak, Y., R. C. Wagenaar, and K. G. Holt, "Gait Characteristics of Elderly People with a History of Falls: A Dynamic Approach," *Physical Therapy*, Vol. 86, No. 11, 2006, pp. 1501–1510.

[160]   Gurbuz, S., et al., "Micro-Doppler Based In-Home Aided and Un-Aided Walking Recognition with Multiple Radar and Sonar Systems," *IET Radar, Sonar & Navigation*, Vol. 11, No. 1, 2016, pp. 107–115.

[161]   Sume, A., et al., "See-Around-Corners with Coherent Radar," *Proceedings of 29th RVK*, Växjö, Sweden, June 9–11, 2008, pp. 217–221.

[162]   Rabaste, O., et al., "Around-The-Corner Radar: Detection of a Human Being in Non-Line of Sight," *IET Radar, Sonar and Navigation*, Vol. 9, No. 6, 2015, pp. 660–668.

[163]   Gustafsson, M., et al., "Extraction of Human Micro-Doppler Signature in an Urban Environment Using a 'Sensing-Behind-the-Corner' Radar," *IEEE Geoscience and Remote Sensing Letters*, Vol. 13, No. 2, 2016, pp. 187–191.

2

Basics of the Micro-Doppler Effect in Radar

The micro-Doppler effect induced by the micromotion of an object or structures on the object can be formulated from both physical and mathematical points of view. The physics of the micro-Doppler effect is derived directly from the theory of the electromagnetic (EM) scattering field. The mathematics of the micro-Doppler effect is derived by introducing micromotion to the conventional Doppler analysis.

Before analyzing radar scattering from an object with micromotion, the basic principle of rigid body and nonrigid body motions and EM scattering from a moving object will be introduced.

2.1   Rigid Body Motion

An object can be a rigid body or a nonrigid body. A rigid body is a solid body with a finite size, but without deformation (i.e., the distance between any two particles of the body does not vary during any motion). This is an idealization, but it efficiently simplifies simulation and analysis.

The mass of a rigid body, M, is the sum of its particle masses, $M = \Sigma_k m_k$, where m_k is the mass of the kth particle. The general motion of a rigid body

51

is a combination of translations (i.e., the parallel motion of all particles in the body) and rotations (i.e., the circular motion of all particles in the body about an axis) [1–3].

To describe the motion of a rigid body, two coordinate systems are commonly used as shown in Figure 2.1: the global or space-fixed system (X, Y, Z) and the local or body-fixed system (x, y, z), which is rigidly fixed in the body. The range vector $\boldsymbol{R}$ is from the origin of the space-fixed system to the origin of the body-fixed system. Let the origin of the body-fixed system be the center of mass (CM) of the body. Then the orientation of the axes in the body-fixed system relative to the axes in the space-fixed system is given by three independent angles. Therefore, the rigid body becomes a mechanical system with 6 degrees of freedom. Let $\boldsymbol{r}$ denote the position of an arbitrary particle P in the body-fixed system. Then its position in the space-fixed system is given by $\boldsymbol{r} + \boldsymbol{R}$, and its velocity is

$$v = \frac{d}{dt}(\boldsymbol{R} + \boldsymbol{r}) = \boldsymbol{V} + \boldsymbol{\Omega} \times \boldsymbol{r} \tag{2.1}$$

where $\boldsymbol{V}$ is the translation velocity of the CM of the rigid body and $\boldsymbol{\Omega}$ is the angular velocity of the body rotation. The direction of $\boldsymbol{\Omega}$ is along the axis of rotation. Thus, a rigid body motion consists of a bulk translational motion and the rotational and/or vibration of the body. As defined in Chapter 1, the rotation and vibration of a body can also be called the micromotion of the body.

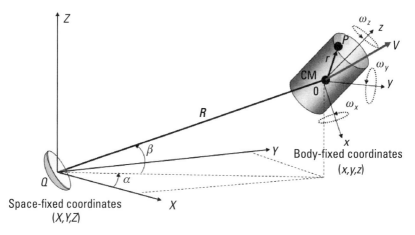

Figure 2.1   Two coordinate systems: the space-fixed system (X, Y, Z) and the body-fixed system (x, y, z) used to describe motion of an object.

To represent the orientation of an object, Euler angles, rotation matrices, and quaternions are the commonly used representations.

2.1.1 Euler Angles

In a rigid body, the rotation about an axis can be described by the rotation axis and the rotation angle using a vector of angular velocity $\mathbf{\Omega}$. The direction of the vector is along the rotation axis. The rotation about an axis can also be described by three rotations about coordinate axes. Euler's rotation theorem states that any two independent orthonormal coordinates are related by a sequence of rotations about coordinate axes [1–4]. There are 12 different sequences available to represent an orientation with Euler angles. They are *x-y-z*, *x-z-y*, *x-y-x*, *x-z-x*, *y-x-z*, *y-z-x*, *y-x-y*, *y-z-y*, *z-x-y*, *z-y-x*, *z-x-z*, and *z-y-z*. The sequence may use the same axis twice but not successively. The order in a rotation sequence is important because the matrix multiplication is not commutative.

The rotation angles (φ, θ, ψ) are called the Euler angles, where φ is defined as the counterclockwise rotation about the *z*-axis, θ is defined as the counterclockwise rotation about the *y*-axis, and ψ is defined as the counterclockwise rotation about the *x*-axis. Euler angles are commonly used to represent three successive rotations in a given rotation sequence. The first step of the rotation sequence rotates the coordinates (x, y, z) to a new coordinates (x_1, y_1, z_1). The second step changes the new coordinates to (x_2, y_2, z_2), and the third step transforms these coordinates to the final coordinates (x_3, y_3, z_3), as illustrated in Figure 2.2.

Some conventions regarding successive rotation sequence are most commonly used. In aerospace engineering, to describe a flying aircraft heading

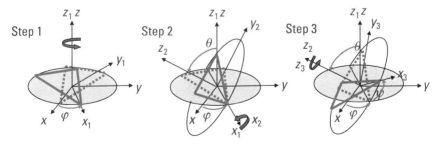

Figure 2.2 Euler angles commonly used to represent three successive rotations.

along the direction of the *x*-axis, with its left side toward the *y*-axis and the upper side to the *z*-axis, three angles rotating about the *x*-axis, *y*-axis, and *z*-axis, called the roll-pitch-yaw convention, are commonly used as illustrated in Figure 2.3. Pitch or attitude is defined by the rotation of θ between $-\pi/2$ and $\pi/2$ about the *y*-axis from the pilot's right side toward the left side, and, thus, the nose of the aircraft pitches up or down. Roll or bank is defined by the rotation of ψ between $-\pi$ and π about the longitudinal *x*-axis of the aircraft from the aircraft tail to its nose. Yaw or heading is defined by the rotation of φ between $-\pi$ and π about the vertical *z*-axis of the aircraft from the bottom toward the top of the aircraft and perpendicular to the other two axes.

Another commonly used rotation sequence is called the *x*-convention in classical mechanics. It follows the *z-x-z* sequence that takes the first rotation by an angle about the *z*-axis, the second rotation by an angle about the *x*-axis, and the third rotation by an angle about the *z*-axis again.

Giving a specific rotation sequence, the rotation matrix is a useful tool for calculating rigid body rotations. Any rigid body specified by its orientation and rotation can be described by its rotation matrix, which can be represented by a product of three elemental rotations defined by the three rotation angles.

For the roll-pitch-yaw or *x-y-z* sequence, the rotation is in the roll-pitch-yaw (ψ-θ-φ) sequence. The first step is rotating about the *x*-axis $x = [1 \;\; 0 \;\; 0]^T$ by an angle ψ defined by the elemental rotation matrix:

$$\Re_X = \begin{bmatrix} 1 & 0 & 0 \\ 0 & \cos\psi & \sin\psi \\ 0 & -\sin\psi & \cos\psi \end{bmatrix} \qquad (2.2)$$

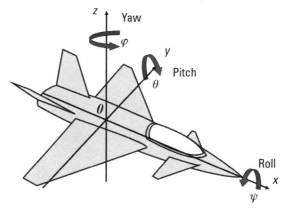

Figure 2.3   The roll-pitch-yaw convention used to describe a flying aircraft.

The second step is rotating about the new y-axis: $y_1 = [0 \quad \cos\psi \quad \sin\psi]^T$ by an angle θ defined by the elemental rotation matrix:

$$\mathfrak{R}_Y = \begin{bmatrix} \cos\theta & 0 & -\sin\theta \\ 0 & 1 & 0 \\ \sin\theta & 0 & \cos\theta \end{bmatrix} \tag{2.3}$$

The third step is rotating about the new z-axis: $z_2 = [-\sin\theta \quad \cos\theta\sin\psi \quad \cos\theta\cos\psi]^T$ by an angle φ defined by the elemental rotation matrix:

$$\mathfrak{R}_Z = \begin{bmatrix} \cos\varphi & \sin\varphi & 0 \\ -\sin\varphi & \cos\varphi & 0 \\ 0 & 0 & 1 \end{bmatrix} \tag{2.4}$$

Thus, the rotation matrix of the roll-pitch-yaw sequence is

$$\mathfrak{R}_{X-Y-Z} = \mathfrak{R}_Z \cdot (\mathfrak{R}_Y \cdot \mathfrak{R}_X) = \begin{bmatrix} r_{11} & r_{12} & r_{13} \\ r_{21} & r_{22} & r_{23} \\ r_{31} & r_{32} & r_{33} \end{bmatrix} \tag{2.5}$$

where the components of the rotation matrix are

$$\begin{cases} r_{11} = \cos\theta\cos\varphi \\ r_{12} = \sin\psi\sin\theta\cos\varphi + \cos\psi\sin\varphi \\ r_{13} = -\cos\psi\sin\theta\cos\varphi + \sin\psi\sin\varphi \end{cases} \tag{2.6}$$

$$\begin{cases} r_{21} = -\cos\theta\sin\varphi \\ r_{22} = -\sin\psi\sin\theta\sin\varphi + \cos\psi\cos\varphi \\ r_{23} = \cos\psi\sin\theta\sin\varphi + \sin\psi\cos\varphi \end{cases} \tag{2.7}$$

$$\begin{cases} r_{31} = \sin\theta \\ r_{32} = -\sin\psi\cos\theta \\ r_{33} = \cos\psi\cos\theta \end{cases} \tag{2.8}$$

From the components of the rotation matrix, if $\cos\theta \neq 0$ or $|r_{11}| + |r_{12}| \neq 0$, three rotation angles (ψ, θ, φ) can be determined as the following:

$$\begin{cases} \psi = \tan^{-1}\left(-r_{32}/r_{33}\right) \\ \theta = \sin^{-1}\left(r_{31}\right) \\ \varphi = \tan^{-1}\left(-r_{21}/r_{11}\right) \end{cases} \qquad (2.9)$$

If $\theta = \pi/2$ or $-\pi/2$, $\cos\theta = 0$, the gimbal lock phenomenon, which will be discussed later, occurs [4–7].

For the x-convention with the z-x-z sequence, the first step is rotating about the z-axis $z = [0\ \ 0\ \ 1]^T$ by an angle φ defined by the elemental rotation matrix R_Z. The second step is rotating about the new x-axis: $x_1 = [\cos\varphi\ \ -\sin\varphi\ \ 0]^T$ by an angle θ defined by the elemental rotation matrix R_X. The third step is rotating about the new z-axis: $z_2 = [\cos\varphi\cos\theta\ \ \sin\varphi\cos\theta\ \ -\sin\theta]^T$ by an angle ψ defined by the elemental rotation matrix R_Z again. Then the rotation matrix of the z-x-z sequence is

$$\Re_{Z-X-Z} = \Re_Z \cdot \left(\Re_X \cdot \Re_Z\right) = \begin{bmatrix} r_{11} & r_{12} & r_{13} \\ r_{21} & r_{22} & r_{23} \\ r_{31} & r_{32} & r_{33} \end{bmatrix} \qquad (2.10)$$

where the components of the rotation matrix are

$$\begin{cases} r_{11} = -\sin\varphi\cos\theta\sin\psi + \cos\varphi\cos\psi \\ r_{21} = -\cos\varphi\cos\theta\sin\psi - \sin\varphi\cos\psi \\ r_{31} = \sin\theta\sin\psi \end{cases} \qquad (2.11)$$

$$\begin{cases} r_{12} = \sin\varphi\cos\theta\cos\psi + \cos\varphi\sin\psi \\ r_{22} = \cos\varphi\cos\theta\cos\psi - \sin\varphi\sin\psi \\ r_{32} = -\sin\theta\cos\psi \end{cases} \qquad (2.12)$$

$$\begin{cases} r_{13} = \sin\varphi\sin\theta \\ r_{23} = \cos\varphi\sin\theta \\ r_{33} = \cos\theta \end{cases} \qquad (2.13)$$

If $\sin\theta \neq 0$ or $|r_{13}| + |r_{23}| \neq 0$, the three rotation angles can be determined from the components of the matrix as the following:

$$\begin{cases} \varphi = \tan^{-1}\left(r_{13}/r_{23}\right) \\ \theta = \cos^{-1}\left(r_{33}\right) \\ \psi = \tan^{-1}\left(r_{31}/-r_{32}\right) \end{cases} \tag{2.14}$$

The rotation matrix $\mathfrak{R}$ is a 3-by-3 matrix and must satisfy the conditions that the product of the matrix and its transposed matrix is a 3-by-3 unit matrix, I, and the determinant of the rotation matrix is +1:

$$\begin{cases} \mathfrak{R}^T\mathfrak{R} = I \\ \det\mathfrak{R} = +1 \end{cases}$$

This means that the three-column vector of the rotation matrix must be orthonormal.

Although the Euler angle rotation matrices are mathematically simpler to handle and easier to understand, they have a problem called gimbal lock [4–7]. Gimbal lock happens when two of the coordinate axes align to each other. It results in a temporary loss of 1 degree of freedom. This phenomenon happens on Earth when an object is at the North Pole or at the South Pole. For example, in the *x-y-z* sequence, when the rotation angle about the *y*-axis (pitch angle) $\theta = \pi/2$, the *x*-axis and *z*-axis will collapse onto one another, and therefore, one degree of freedom will be lost. Recall (2.5) through (2.8), when the rotation angle about the *y*-axis (pitch angle) $\theta = \pi/2$, the rotation matrix becomes

$$\mathfrak{R}_{X-Y-Z}\left(\psi,\theta = \frac{\pi}{2},\varphi\right) = \begin{bmatrix} 0 & 0 & -1 \\ \sin\psi\cos\varphi - \cos\psi\sin\varphi & \sin\psi\sin\varphi + \cos\psi\cos\varphi & 0 \\ \cos\psi\cos\varphi + \cos\psi\cos\varphi & \cos\psi\sin\varphi - \sin\psi\cos\varphi & 0 \end{bmatrix} \tag{2.15}$$

If $\psi = 0$, the rotation matrix is

$$\mathfrak{R}_{X-Y-Z}\left(\psi = 0,\theta = \frac{\pi}{2},\varphi\right) = \begin{bmatrix} 0 & 0 & -1 \\ -\sin\varphi & \cos\varphi & 0 \\ \cos\varphi & \sin\varphi & 0 \end{bmatrix} \tag{2.16}$$

When $\varphi = 0$ and $\psi = -\psi$, the rotation matrix is

$$\mathfrak{R}_{X-Y-Z}\left(\psi = -\psi,\theta = \frac{\pi}{2},\varphi = 0\right) = \begin{bmatrix} 0 & 0 & -1 \\ -\sin\psi & \cos\psi & 0 \\ \cos\psi & \sin\psi & 0 \end{bmatrix} \tag{2.17}$$

Thus, the value of the 3-by-3 rotation matrix $\Re_{X-Y-Z}(0, \pi/2, \varphi)$ equals to the value of the rotation matrix $\Re_{X-Y-Z}(-\psi, \pi/2, 0)$, and one degree of freedom is lost. In this case, there are only two degrees of freedom. Thus, the gimbal lock makes unexpected loss of one degree of freedom and limits the movement.

However, an alternative that uses quaternion algebra instead of the Euler angle rotation matrices can eliminate the gimbal lock. Quaternions are commonly used in computer graphics for representing orientation in three-dimensional (3-D) space. Instead of using three rotation angles, the quaternion represents an arbitrary orientation by a rotation about a unit axis through a certain angle.

2.1.2 Quaternion

The problem associated with the Euler rotation matrices comes from the trigonometric function operation. When an Euler angle reaches $\pm\pi/2$, a numerical singularity may occur. Quaternion algebra alleviates the singularity by using rotation around an axis instead of using the yaw, pitch, and roll angular rotation [4–7]. By using the quaternion, there is only a single axis of rotation and it does not require predefined Euler angle sequences. Thus, no degrees of freedom are lost.

A quaternion uses a four-component vector to represent 3-D orientation. The real part is a scalar rotation angle α, and the imaginary part is a vector with x, y, z coordinates. If a point at (x_1, y_1, z_1) in a 3-D space $P_1 = x_1 \cdot i + y_1 \cdot j + z_1 \cdot k$ rotates through an angle α about a unit axis $e = q_1 \cdot i + q_2 \cdot j + q_3 \cdot k = (q_1, q_2, q_3)$, the resulting point can be represented by the following transformation formula [6]:

$$P_2 = q \cdot P_1 \cdot \text{conj}(q) \tag{2.18}$$

where $q(\alpha, e) = [\cos(\alpha/2), e \cdot \sin(\alpha/2)]$ is the quaternion representing the rotation about the unit vector e by an angle α. A translation can be represented by $P_2 = q + P_1$.

The quaternion can also be represented by four components as

$$q = q_0 u + q_1 i + q_2 j + q_3 k = \left[q_0, q_1, q_2, q_3 \right]$$
$$= \left[\cos\left(\frac{\alpha}{2}\right), x\sin\left(\frac{\alpha}{2}\right), y\sin\left(\frac{\alpha}{2}\right), z\sin\left(\frac{\alpha}{2}\right) \right] \tag{2.19}$$

where $q_0 = \cos(\alpha/2)$, $q_1 = x\sin(\alpha/2)$, $q_2 = y\sin(\alpha/2)$, and $q_3 = z\sin(\alpha/2)$ with the constraint:

$$q_0^2 + q_1^2 + q_2^2 + q_3^2 = 1$$

The conjugate of a quaternion, $conj(q)$, is a quaternion that has the same magnitudes, but the signs of the imaginary parts are changed:

$$\begin{aligned} \text{conj}(\boldsymbol{q}) &= \left[q_0, -q_1, -q_2, -q_3 \right] \\ &= \left[\cos\left(\frac{\alpha}{2}\right), -x\sin\left(\frac{\alpha}{2}\right), -y\sin\left(\frac{\alpha}{2}\right), -z\sin\left(\frac{\alpha}{2}\right) \right] \end{aligned} \quad (2.20)$$

Multiplying a quaternion by its conjugate gives a real number: $\boldsymbol{q} \cdot conj(\boldsymbol{q})$ = real number. The norm of a quaternion is defined by

$$|\boldsymbol{q}| = \left[\boldsymbol{q} \cdot \text{conj}(\boldsymbol{q}) \right]^{1/2} \quad (2.21)$$

and

$$|\boldsymbol{q}|^2 = \cos^2\left(\frac{\alpha}{2}\right) + q_1^2 + q_2^2 + q_3^2 = \cos^2\left(\frac{\alpha}{2}\right) + S^2 \quad (2.22)$$

where $S = (q_1^2 + q_2^2 + q_3^2)^{1/2}$. The normalized quaternion is defined by $|\boldsymbol{q}| = 1$.

The multiplication of two quaternions, $\boldsymbol{p} = [p_0, p_1, p_2, p_3]$ and $\boldsymbol{q} = [q_0, q_1, q_2, q_3]$, is given by [4]

$$\begin{aligned} \boldsymbol{p} \cdot \boldsymbol{q} = {}& p_0 q_0 - \left(p_1 q_1 + p_2 q_2 + p_3 q_3 \right) \\ & + p_0\left(q_1 \cdot \boldsymbol{i} + q_2 \cdot \boldsymbol{j} + q_3 \cdot \boldsymbol{k} \right) + q_0\left(p_1 \cdot \boldsymbol{i} + p_2 \cdot \boldsymbol{j} + p_3 \cdot \boldsymbol{k} \right) \\ & + \left(p_2 q_3 - p_3 q_2 \right) \cdot \boldsymbol{i} + \left(p_3 q_1 - p_1 q_3 \right) \cdot \boldsymbol{j} + \left(p_1 q_2 - p_2 q_1 \right) \cdot \boldsymbol{k} \end{aligned} \quad (2.23)$$

It should be noted that quaternions are not commutative under the multiplication operation, that is, $\boldsymbol{p} \cdot \boldsymbol{q} \neq \boldsymbol{q} \cdot \boldsymbol{p}$.

Given a rotation angle α and a unit axis $\boldsymbol{e} = (q_1 \cdot \boldsymbol{i}, q_2 \cdot \boldsymbol{j}, q_3 \cdot \boldsymbol{k})$, the unit vector in (x, y, z) is defined by $\boldsymbol{u} = \boldsymbol{e}/S = [x \cdot \boldsymbol{i}, y \cdot \boldsymbol{j}, z \cdot \boldsymbol{k}] = [(q_1/S)\boldsymbol{i}, (q_2/S)\boldsymbol{j}, (q_3/S)\boldsymbol{k}]$, and the corresponding rotation matrix can be derived by the quaternion $\boldsymbol{q}(\alpha, \boldsymbol{e}) = [\cos(\alpha/2), \boldsymbol{e} \cdot \sin(\alpha/2)]$ [6]:

$$\Re(\alpha,e) =$$

$$\begin{bmatrix} q_1^2 + \cos\alpha\left(1 - q_1^2\right) & q_1 q_2 (1 - \cos\alpha) + q_3 \sin\alpha & q_1 q_3 (1 - \cos\alpha) - q_2 \sin\alpha \\ q_1 q_2 (1 - \cos\alpha) - q_3 \sin\alpha & q_2^2 + \cos\alpha\left(1 - q_2^2\right) & q_2 q_3 (1 - \cos\alpha) + q_1 \sin\alpha \\ q_1 q_3 (1 - \cos\alpha) + q_2 \sin\alpha & q_2 q_3 (1 - \cos\alpha) - q_1 \sin\alpha & q_3^2 + \cos\alpha\left(1 - q_3^2\right) \end{bmatrix}$$

$$(2.24)$$

or represented by $\boldsymbol{q} = [q_0, q_1, q_2, q_3]$ as

$$\Re(q_0, q_1, q_2, q_3) = \begin{bmatrix} 1 - 2q_2^2 - 2q_3^2 & 2q_1 q_2 - 2q_0 q_3 & 2q_1 q_3 + 2q_0 q_2 \\ 2q_1 q_2 + 2q_0 q_3 & 1 - 2q_1^2 - 2q_3^2 & 2q_2 q_3 - 2q_0 q_1 \\ 2q_1 q_3 - 2q_0 q_2 & 2q_2 q_3 + 2q_0 q_1 & 1 - 2q_1^2 - 2q_2^2 \end{bmatrix} \quad (2.25)$$

In some applications, if the Euler angles are already given, they can be used along with quaternions to take the advantage of the quaternion. Euler angles can be easily converted to quaternions. Any of Euler angle conventions can be performed with quaternions. Using quaternion components, each roll, pitch, and yaw rotation is described by $q_{\text{roll}} = [\cos(\psi/2), \sin(\psi/2), 0, 0]$, $q_{\text{pitch}} = [\cos(\vartheta/2), 0, \sin(\vartheta/2), 0]$, and $q_{\text{yaw}} = [\cos(\varphi/2), 0, 0, \sin(\varphi/2)]$.

With the x-y-z sequence, given the roll, pitch and yaw angles (ψ, θ, φ), the quaternion is [6]:

$$\boldsymbol{q}_{x-y-z}(\psi,\theta,\varphi) = \left[q_0, q_1, q_2, q_3 \right] \quad (2.26)$$

where

$$\begin{aligned} q_0 &= \cos\frac{\psi}{2}\cos\frac{\theta}{2}\cos\frac{\varphi}{2} + \sin\frac{\psi}{2}\sin\frac{\theta}{2}\sin\frac{\varphi}{2}; \\ q_1 &= \sin\frac{\psi}{2}\cos\frac{\theta}{2}\cos\frac{\varphi}{2} - \cos\frac{\psi}{2}\sin\frac{\theta}{2}\sin\frac{\varphi}{2}; \\ q_2 &= \cos\frac{\psi}{2}\sin\frac{\theta}{2}\cos\frac{\varphi}{2} + \sin\frac{\psi}{2}\cos\frac{\theta}{2}\sin\frac{\varphi}{2}; \\ q_3 &= \cos\frac{\psi}{2}\cos\frac{\theta}{2}\sin\frac{\varphi}{2} - \sin\frac{\psi}{2}\sin\frac{\theta}{2}\cos\frac{\varphi}{2}. \end{aligned} \quad (2.27)$$

The quaternion can also be converted to Euler angles by first converting the quaternion to a matrix and then converting the matrix to Euler angles.

However, if the quaternion combined with Euler angles is not used properly, the gimbal lock can still occur because of the use of three rotations.

2.1.3 Equations of Motion

A rigid body in motion is described by its kinematics and dynamics [1, 3]. The kinematics of rigid body motion describes the relation between the position, speed, and acceleration of the body motion without considering what forces cause the motion. The dynamics or kinetics of rigid body motion describes motion of the body and the forces that affect the motion. The dynamic energy of a body is given by [1–3]

$$E = \frac{1}{2}\sum_k m_k v_k^2 = \frac{1}{2}\sum_k m_k \left(V + \Omega \times r_k\right)^2$$

$$= \frac{1}{2}MV^2 + \frac{1}{2}\sum_k \left[\Omega^2 r_k^2 - \left(\Omega \cdot r_k\right)^2\right] = E_{CM} + E_{\mathrm{Rot}} \qquad (2.28)$$

where m_k, v_k, and r_k denote the mass, velocity, and the position of a point k in the body-fixed system where $\Sigma_k m_k r_k = 0$ is assumed, the $V = \| V \|$ is the norm of the translation velocity of the rigid body, and $\Omega = \| \Omega \|$ is the norm of the angular velocity of the body rotation. Because of the constant V, $M = \Sigma_k m_k$, the dynamic energy of a rigid body is the sum of the dynamic energy of the translational motion of the center of mass (CM), the E_{CM}, and the dynamic energy of the rigid body rotation about the CM, the E_{Rot}.

In rigid body motion, rotation plays a critical role. Given the angular momentum about the CM of a rigid body defined by

$$L = \sum_k m_k r_k \times \left(\Omega \times r_k\right) = \sum_k m_k \left[r_k^2 \Omega - r_k \left(r_k \cdot \Omega\right)\right] \qquad (2.29)$$

the dynamic energy of the rotating body can be expressed as

$$E_{\mathrm{Rot}} = \frac{1}{2}\sum_{i,j} \Omega_i \Omega_j \sum_k m_k \left[\sum_l \left(r_k\right)_l^2 \delta_{i,j} - \left(r_k\right)_i \left(r_k\right)_j\right] = \frac{1}{2}\sum_{i,j} I_{i,j} \Omega_i \Omega_j \qquad (2.30)$$

where Ω_j is the component of Ω along the j-axis of the body-fixed system, r_k is expressed by $[(r_k)_1, (r_k)_2, (r_k)_3]$, and $I_{i,j}$ is the inertia tensor defined by

$$I_{i,j} = \frac{1}{2} \sum_k m_k \left[\sum_l (r_k)_l^2 \delta_{i,j} - (r_k)_i (r_k)_j \right] \qquad (2.31)$$

Thus, the component angular momentum becomes

$$L_i = \sum_j I_{i,j} \Omega_j = \sum_j \sum_k m_k \left[\sum_l (r_k)_l^2 \delta_{i,j} - (r_k)_i (r_k)_j \right] \Omega_j \qquad (2.32)$$

By appropriate choice of the orientation of the body-fixed coordinate system, the moment inertia tensor can be reduced to a diagonal form. In this case, the directions of the axes are called the principal axes of the moment inertia tensor, and the diagonal components of the tensor are called the principal moments of inertia. Then the kinetic energy of the rotating body becomes

$$E_{\mathrm{Rot}} = \frac{1}{2} \left(I_1 \Omega_1^2 + I_2 \Omega_2^2 + I_3 \Omega_3^2 \right) \qquad (2.33)$$

and the angular momentum vector L becomes

$$L = L_1 e_1 + L_2 e_2 + L_3 e_3 = I_1 \Omega_1 e_1 + I_2 \Omega_2 e_2 + I_3 \Omega_3 e_3 \qquad (2.34)$$

where $[e_1, e_2, e_3]$ is the unit vector in the directions of the principal axes, $[\Omega_1, \Omega_2, \Omega_3]$ and $[L_1, L_2, L_3]$ are the angular velocity vector and the angular momentum vector along the principal axes, respectively.

The Euler equation of motion is derived by taking the time derivative of the angular momentum. In the rotating body-fixed coordinates, the time derivative of the angular momentum is replaced by

$$\left(\frac{d}{dt} L \right)_{\mathrm{Rot}} + \Omega \times L = F \qquad (2.35)$$

In the space-fixed coordinates, the time derivative of the angular momentum equals the applied external force F:

$$\frac{d}{dt} L = \frac{d}{dt} (I \cdot \Omega) = F \qquad (2.36)$$

Therefore, the Euler equations of motion become the following set of differential equations [3]:

$$I_1 \frac{d\Omega_1}{dt} + \left(I_3 - I_2\right)\Omega_2\Omega_3 = F_1;$$

$$I_2 \frac{d\Omega_2}{dt} + \left(I_1 - I_3\right)\Omega_3\Omega_1 = F_2; \qquad (2.37)$$

$$I_3 \frac{d\Omega_3}{dt} + \left(I_2 - I_1\right)\Omega_1\Omega_2 = F_3.$$

These differential equations that govern the motion will be used in calculating rigid body nonlinear motion dynamics.

2.2 Nonrigid Body Motion

The nonrigid body is a deformable body, that is, a force acts on the body that will lead to the body changing its shape, called deformation. For a free-form deformation, each particle in the body moves from an initial position P to a new position $P_t = f(t, P)$ at time t, which is a function of t and P. Stress and strain may occur everywhere in the body. To compute the deformation of a complex body, more complicated methods such as the finite difference method (FDM), the finite element method (FEM), and the boundary element method may be used [8].

In this book, any nonrigid body motion is modeled by jointly connected rigid body segments or parts [2, 8–10]. In robotics, the robot arm is considered to be a flexible part of a mechanical system. The joint connection of two rigid segments of the arm is defined by the kinematical constraint on the joint that restricts the relative motions of the two individual rigid segments. When studying radar scattering from a nonrigid body motion, the nonrigid body (such as a walking human or a flying bird) is modeled as several jointly connected rigid body segments. The motion of each segment is treated as a rigid body motion. In mechanics, this type of nonrigid body system is defined as a multibody system.

The multibody system is widely used to model, simulate, analyze, and optimize motion kinematics and dynamic behavior of interconnected bodies in robotics and vehicle dynamics [8–10]. An example of the simplest multibody system is called the slider-crank mechanism. The mechanism is used to transform the circular rotation of the crank into a linear translation of the piston; alternately, when the piston is forced to move, the linear translation is converted into circular rotation. The slider is not allowed to rotate and three revolute joints are used to connect these interconnected bodies.

The crank, connecting-rod, and piston mechanism is shown in Figure 2.4. The kinematic analysis of the slider-crank mechanism gives the displacement (or position), velocity, and acceleration of the piston or the connecting rod. From the geometry, the piston displacement x is determined by the length of the connecting rod L, the length of the crank R, the crank angle θ, and the connecting-rod angle φ

$$x = R\cos\theta + L\cos\varphi \qquad (2.38)$$

Since the connecting-rod angle φ and the crank angle θ are related by

$$R\sin\theta = -L\sin\varphi \qquad (2.39)$$

or

$$\sin\varphi = -\frac{R}{L}\sin\theta \qquad (2.40)$$

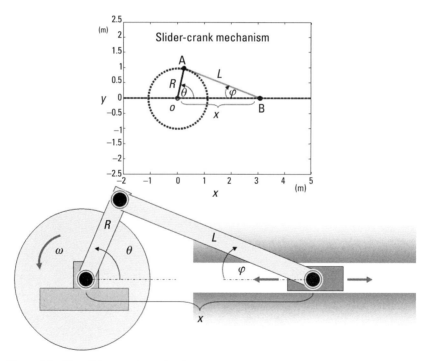

Figure 2.4   The slider-crank mechanism.

and

$$\cos\varphi = \left(1 - \frac{R^2}{L^2}\sin^2\theta\right)^{1/2} \tag{2.41}$$

(2.38) can be rewritten as

$$x = R\cos\theta + L\cos\varphi = R\cos\theta + L\left(1 - \sin^2\varphi\right)^{1/2} = R\cos\theta + \left(L^2 - R^2\sin^2\theta\right)^{1/2} \tag{2.42}$$

Given the angular velocity of the rotating crank $d\theta/dt = \Omega$, by taking the time derivative of both sides of (2.38),

$$R\cos\theta\frac{d\theta}{dt} = -L\cos\varphi\frac{d\varphi}{dt} \tag{2.43}$$

or

$$\frac{d\varphi}{dt} = -\frac{R\cos\theta}{L\cos\varphi}\cdot\Omega \tag{2.44}$$

Thus, the angular velocity of the connecting rod is

$$\frac{d\varphi}{dt} = -\frac{R\cos\theta}{\left(L^2 - R^2\sin^2\theta\right)^{1/2}}\cdot\Omega \tag{2.45}$$

The translational velocity of the slider becomes

$$\frac{dx}{dt} = -R\cdot\sin\theta\cdot\Omega - \frac{R^2\cdot\sin\theta\cdot\cos\theta}{\left(L^2 - R^2\cdot\sin^2\theta\right)^{1/2}}\cdot\Omega \tag{2.46}$$

The angular acceleration of the connecting rod and the slider translational acceleration are

$$\frac{d^2\varphi}{dt^2} = \frac{R\cdot\sin\theta\cdot\Omega^2 + L\cdot\sin\varphi\cdot\left(\dfrac{d\varphi}{dt}\right)^2}{L\cdot\cos\varphi} \tag{2.47}$$

and

$$\frac{d^2x}{dt^2} = -R \cdot \cos\theta \cdot \Omega^2 - L \cdot \cos\varphi \cdot \left(\frac{d\varphi}{dt}\right)^2 - L \cdot \sin\varphi \cdot \frac{d^2\varphi}{dt^2} \quad (2.48)$$

respectively.

Given $R = 1.0$m, $L = 3.0$m, and $\Omega = 2\pi$ (rad/sec), Figure 2.5 shows the variations in displacement, velocity and acceleration of the piston as a function of time, and the variations in the connecting-rod angle, angular velocity, and angular acceleration as a function of time.

Based on the kinematic analysis, a robot may be modeled by a multibody system to simulate motion kinematics and dynamic behavior of the interconnected body segments. Similarly, a human body can also be modeled by a multibody system, and each interconnected body segment is considered a rigid body.

2.3   Electromagnetic Scattering from a Body with Motion

Before introducing the physics of the micro-Doppler effect in radar and calculating radar EM scattering from a target, the basics of EM scattering should be briefly introduced.

The simplest EM scattering model for a target is the point scatterer model. The target is defined in terms of a 3-D reflectivity density function

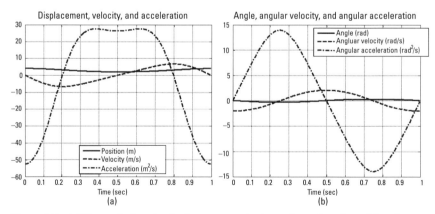

Figure 2.5   (a) The variations in displacement, velocity and acceleration of the piston as a function of time, and (b) the variations in the connecting-rod angle, angular velocity, and angular acceleration as a function of time.

characterized by point scatterers. The occlusion effect can also be implemented in the point scatterer model. Compared with other EM scattering models, the point scatterer model can easily incorporate the target's motion in EM scattering and isolate the EM scattering from each individual motion component.

2.3.1 Radar Cross Section of a Target

EM scattering occurs when a target is illuminated by radar-transmitted EM waves. The incident EM waves induce electric and magnetic currents on the surface and/or within the volume of the target that will generate a scattered EM field. The scattered EM waves are transmitted to all possible directions. If the target is at a distance far enough from the radar, the incident wavefront can be treated as a plane wave. The power of scattered EM waves is measured by a bistatic scattering cross section of the target. If the direction is back to the radar, the bistatic scattering becomes backscattering and the cross section is a backscattering cross-section, called the radar cross-section (RCS).

According to [11], "the *IEEE Dictionary of Electrical and Electronics Terms* defines RCS as a measure of reflective strength of a target defined as 4π times the ratio of the power per unit solid angle scattered in a specified direction to the power per unit area in a plane wave incident on the scatterer from a specified direction." The RCS is formulated by $\sigma = \lim_{r\to\infty} 4\pi r^2 |E_s|^2 / |E_i|^2$, where E_s is the intensity of the far field scattered electric field, E_i is the intensity of the far-field incident electric field, and r is the distance from the radar to the target.

The RCS is defined to characterize the target characteristics. It is normalized to the power density of the incident wave at the target in that it does not depend on the distance of the target from the radar. The RCS is dependent on the size, geometry, and material of the target, the frequency of the transmitter, the polarization of the transmitter and the receiver, and the aspect angles of the target relative to the radar transmitter and the receiver, respectively [11–13].

The maximum detectable range of a target is proportional to the fourth root of its RCS. The RCS is usually specified in units of square meters (m²). Typical RCS ranges are from 0.0005 m² for insects, 0.01 m² for small birds, 0.5 m² for humans, and up to 100,000 m² for a large ship.

Any complicated target can be decomposed into basic geometric building blocks of simple shapes such as spheres, cylinders, and plates. Figure 2.6 illustrates an example of a human modeled by simple triangle-shaped surface geometric building blocks.

When a target is illuminated by EM waves, each building block produces a voltage. The vector sum of the building block voltages determines the total

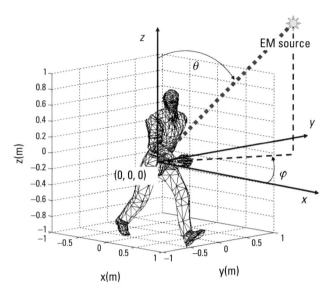

Figure 2.6   A human modeled by simple triangle shaped surface building blocks.

RCS of the target. It is defined by the square root of the magnitude of the vector sum. The accuracy of the RCS calculation depends on the accuracy of modeling these building blocks and the interactions between them. If the modeling is not accurate, the calculated RCS may not agree with the measured RCS.

The mechanism of the EM scattering from a target is a complicated process that includes reflections, diffractions, surface waves, ducting, and interactions between them. Reflection is from surfaces and has the highest RCS peak of the scattering mechanisms. Diffraction is from discontinuities (such as edges, corners, or vertices) and is less intense than reflection. The surface wave is the current traveling along the surface of the target body. A flat surface produces leaky waves and curved surfaces produce creeping waves. Ducting occurs when a wave enters into a waveguide-like structure (such as the inlet cavity of an aircraft). Spiky features and lobes in the RCS may also associate with multiple reflections, diffraction, and other scattering mechanisms.

2.3.2   RCS Prediction Methods

The RCS prediction method is an analytical method of calculating the RCS. The incident wave induces a current on the target and thus radiates an EM field. If the distribution of current is known, it can be used in the radiation integrals to calculate the scattered field and the RCS. The commonly used methods for RCS prediction are the physical optics (PO), ray tracing, method of moments, and finite-difference methods [13]. The PO method is a high-frequency approximation to estimate the surface current induced on a body. It is accurate in the specular direction, but inaccurate when computing at angles far from the specular directions or in the shadow regions. However, surface waves are not included in the PO method. To improve the accuracy for the current distribution near edges, the physical theory of diffraction (PTD) may be used.

The ray-tracing method is used to analyze large objects with arbitrary shape. Geometric optics (GO) is the classical theory of ray tracing and provides a formula for computing the reflected and refracted fields. It can also be supplemented by the geometric theory of diffraction (GTD). Computer codes based on GTD, PTD, and their hybridizations have been developed for the prediction of high-frequency scattering from complex perfectly conducting objects.

The XPATCH code, based on the shooting-and-bouncing ray technique and PTD, has been widely used to generate a target's RCS signatures for noncooperative target recognition [14]. It allows the calculation of backscattering from complex geometries. Other computer codes for RCS prediction include the RAPPORT code developed in the Netherlands, the numerical electromagnetic code (NEC), the electric field integral equation (EFIE), and the finite-difference time domain (FDTD) [13, 15].

In this book, for some applications, the PO method will be used to estimate scattered fields. A simple RCS prediction code based on the PO method called the POFACET is available [16, 17]; this can calculate the monostatic RCS and the bistatic RCS of a static object. The bistatic RCS determines the EM power flux density scattered by an object in an arbitrary direction. In using the POFACET of the RCS prediction method, an object is usually approximated by a large number of subdivision surfaces (triangular meshes) called facets that produce a continuous surface of the object. The total RCS of the object is the superposition of the square root of the magnitude of each individual facet's RCS. The scattered field of each triangle is computed by assuming that the triangle is isolated and other triangles are not present. Besides multiple reflections, edge diffraction and surface waves are not considered. Shadowing is only approximately included by considering a facet to be completely illuminated or completely shadowed by the incident wave. A standard spherical coordinate system is used in the POFACET RCS prediction to specify incident and scattering directions as shown in Figure 2.7, and the RCS is calculated at specific angles θ and φ.

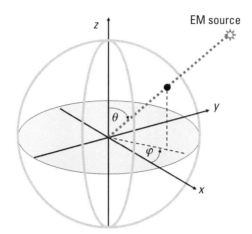

Figure 2.7 Coordinate system used in the RCS calculation.

2.3.3 EM Scattering from a Body with Motion

The characteristics of the EM scattering field from an object have been studied for decades. The characteristic of the scattering field and the RCS of objects are usually calculated under an assumption that the object is stationary. However, in most practical situations, an object or structures on the object are rarely stationary and may have movements such as translation, rotation, or oscillation.

The EM field scattered from a moving object or an oscillating object has been studied in both theory and experiment [18–20]. Theoretical analysis indicates that the motion of an object modulates the phase function of the scattered EM waves. If the object oscillates linearly and periodically, the modulation generates sideband frequencies about the Doppler frequency induced by its translation.

For a translational object, the far electric field of the object can be derived as [19]

$$E_T(r') = \exp\left\{ jkr_0 \cdot (u_k - u_r) \right\} E(r) \tag{2.49}$$

where $k = 2\pi/\lambda$ is the wave number, u_k is the unit vector of the incidence wave, u_r is the unit vector of the direction of observation, $E(r)$ is the far electric field of the object before moving, $r = (X_0, Y_0, Z_0)$ are the initial coordinates of the object in the space-fixed coordinates (X, Y, Z), $r' = (X_1, Y_1, Z_1)$ are the coordinates of the object in the space-fixed coordinates after translation, and $r' = r + r_0$, where r_0 is the translation vector, as illustrated in Figure 2.8.

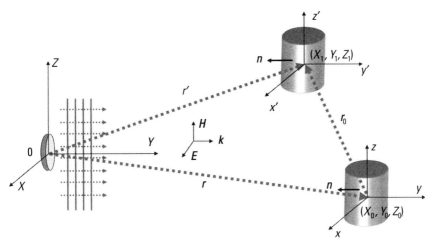

Figure 2.8 Geometry of a translational object in a far EM field.

The only difference in the electric field before and after the translation is the phase term, $\exp\{jk\boldsymbol{r}_0 \cdot (\boldsymbol{u}_k - \boldsymbol{u}_r)\}$. If the translation is a function of time, $\boldsymbol{r}_0 = \boldsymbol{r}_0(t) = r_0(t)\boldsymbol{u}_T$, where $\boldsymbol{u}_T$ is the unit vector of the translation, the phase factor then becomes

$$\exp\{j\Phi(t)\} = \exp\{jkr_0(t)\boldsymbol{u}_T \cdot (\boldsymbol{u}_k - \boldsymbol{u}_r)\} \qquad (2.50)$$

For backscattering, the direction of observation is opposite to the direction of the incidence wave. Thus, $\boldsymbol{u}_k = -\boldsymbol{u}_r$ and

$$\exp\{j\Phi(t)\} = \exp\{j2kr_0(t)\boldsymbol{u}_T \cdot \boldsymbol{u}_k\} \qquad (2.51)$$

If the translation direction is perpendicular to the direction of the incidence wave, that is, $\boldsymbol{u}_T \cdot \boldsymbol{u}_k = 0$, then $\exp\{\Phi(t)\} = 1$.

For a vibrating object, assuming $r_0(t) = A\cos\Omega t$, the phase factor becomes a periodic function of time with an angular frequency Ω

$$\exp\{j\Phi(t)\} = \exp\{j2kA\cos\Omega t\,\boldsymbol{u}_T \cdot \boldsymbol{u}_k\} \qquad (2.52)$$

In general, when radar transmits an EM wave at a carrier frequency f_0, the radar received signal can be expressed as

$$s(t) = \exp\{jkr_0(t)(\boldsymbol{u}_k - \boldsymbol{u}_r) - j2\pi f_0 t\}|E(r)| \qquad (2.53)$$

where the phase factor, $\exp\{jk\boldsymbol{r}_0(t)(\boldsymbol{u}_k - \boldsymbol{u}_r)\}$, defines the modulation of the micro-Doppler effect caused by the time-varying motion $\boldsymbol{r}_0(t)$.

2.4 Basic Mathematics for Calculating the Micro-Doppler Effect

The mathematics of the micro-Doppler can be derived by introducing the micromotion to the conventional Doppler analysis [21]. For simplicity, a radar target is represented by a set of point scatterers that are primary reflecting points on the target. The point scattering model simplifies the analysis while preserving micro-Doppler features. In the simplified model, scatterers are assumed to be perfect reflectors, reflecting all the energy intercepted.

As shown in Figure 2.9, the radar is stationary and located at the origin Q of the radar-fixed coordinate system (X, Y, Z). The target is described in a local coordinate system (x, y, z) that is attached to the target and has translation and rotation with respect to the radar coordinates. To observe the target's rotations, a reference coordinate system (X', Y', Z') is introduced, which shares the same origin with the target local coordinates and thus has the same translation as the target but no rotation with respect to the radar coordinates. The origin O of the reference coordinates is assumed to be at a distance R_0 from the radar.

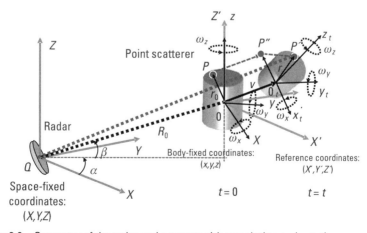

Figure 2.9 Geometry of the radar and a target with translation and rotation.

2.4.1 Micro-Doppler Induced by a Target with Micromotion

Suppose the target has a translation velocity $\boldsymbol{v}$ with respect to the radar and an angular rotation velocity $\boldsymbol{\omega}$, which can be either represented in the target local coordinate system as $\boldsymbol{\omega} = (\omega_x, \omega_y, \omega_z)^T$, or represented in the reference coordinate system as $\boldsymbol{\omega} = (\omega_X, \omega_Y, \omega_Z)^T$. Thus, a point scatterer P at time $t = 0$ will move to P' at time t. The movement consists of two steps: (1) translation from P to P'', as shown in Figure 2.9, with a velocity $\boldsymbol{v}$, that is, $OO_t = \boldsymbol{v}t$; and (2) rotation from P'' to P' with an angular velocity $\boldsymbol{\omega}$. If we observe the movement in the reference coordinate system, the point scatterer P is located at $\boldsymbol{r}_0 = (X_0, Y_0, Z_0)^T$, and the rotation from P'' to P' is described by a rotation matrix $\mathfrak{R}_t$. Then, at time t the location of P' will be at

$$\boldsymbol{r} = O_t P' = \mathfrak{R}_t O_t P'' = \mathfrak{R}_t \boldsymbol{r}_0 \tag{2.54}$$

The range vector from the radar at Q to the scatterer at P' can be derived as

$$QP' = QO + OO_t + O_t P' = \boldsymbol{R}_0 + \boldsymbol{v}t + \mathfrak{R}_t \boldsymbol{r}_0 \tag{2.55}$$

and the scalar range becomes

$$r(t) = \left\| \boldsymbol{R}_0 + \boldsymbol{v}t + \mathfrak{R}_t \boldsymbol{r}_0 \right\| \tag{2.56}$$

where $\|\cdot\|$ represents the Euclidean norm.

If the radar transmits a sinusoidal waveform with a carrier frequency f, the baseband of the signal returned from the point scatterer P is a function of $r(t)$:

$$s(t) = \rho(x, y, z)\exp\left\{ j2\pi f \frac{2r(t)}{c} \right\} = \rho(x, y, z)\exp\left\{ j\Phi[r(t)] \right\} \tag{2.57}$$

where $\rho(x, y, z)$ is the reflectivity function of the point scatterer P described in the target local coordinates (x, y, z), c is the speed of the EM wave propagation, and the phase of the baseband signal is

$$\Phi(r) = 2\pi f \frac{2r(t)}{c} \tag{2.58}$$

By taking the time derivative of the phase, the Doppler frequency shift of the point scatterer P induced by the target's motion can be derived [21]

$$f_D = \frac{1}{2\pi}\frac{d\Phi(t)}{dt} = \frac{2f}{c}\frac{d}{dt}r(t)$$

$$= \frac{2f}{c}\frac{1}{2r(t)}\frac{d}{dt}\left[\left(R_0 + vt + \Re_t r_0\right)^T \left(R_0 + vt + \Re_t r_0\right)\right] \qquad (2.59)$$

$$= \frac{2f}{c}\left[v + \frac{d}{dt}\left(\Re_t r_0\right)\right]^T n$$

where $n = (R_0 + vt + \Re_t r_0)/\parallel R_0 + vt + \Re_t r_0 \parallel$ is the unit vector of QP'.

Before further deriving the Doppler shift induced by the rotation, a useful relationship $u \times r = \hat{u}r$ is introduced. Given a vector $u = [u_x, u_y, u_z]^T$ and a skew symmetric matrix defined by

$$\hat{u} = \begin{bmatrix} 0 & -u_z & u_y \\ u_z & 0 & -u_x \\ -u_y & u_x & 0 \end{bmatrix} \qquad (2.60)$$

the cross-product of the vector u and any vector r can be computed through the matrix computation:

$$u \times r = \begin{bmatrix} u_y r_z - u_z r_y \\ u_z r_x - u_x r_z \\ u_x r_y - u_y r_x \end{bmatrix} = \begin{bmatrix} 0 & -u_z & u_y \\ u_z & 0 & -u_x \\ -u_y & u_x & 0 \end{bmatrix}\begin{bmatrix} r_x \\ r_y \\ r_z \end{bmatrix} = \hat{u}r \qquad (2.61)$$

This equation is useful in the analysis of the special orthogonal matrix group or $SO(3)$ group, also called the 3-D rotation matrix [2, 21].

We can now return to the rotation matrix in (2.59). In the reference coordinate system, the angular rotation velocity vector can be described by $\omega = (\omega_X, \omega_Y, \omega_Z)^T$, and the target will rotate along the unit vector $\omega' = \omega/\parallel\omega\parallel$ with a scalar angular velocity $\Omega = \parallel\omega\parallel$. Assuming a high PRF and a relatively low angular velocity, the rotational motion during each time interval can be considered to be infinitesimal, and thus (see Appendix 2A)

$$\Re_t = \exp\{\hat{\omega}t\} \qquad (2.62)$$

where $\hat{\omega}$ is the skew symmetric matrix associated with ω. Thus, the Doppler frequency shift in (2.59) becomes

$$f_D = \frac{2f}{c}\left[v + \frac{d}{dt}\left(e^{\hat{\omega}t}r_0\right)\right]^T n = \frac{2f}{c}\left(v + \hat{\omega}e^{\hat{\omega}t}r_0\right)^T n$$

$$= \frac{2f}{c}(v + \hat{\omega}\,r)^T n = \frac{2f}{c}(v + \omega \times r)^T n$$

(2.63)

If $\|R_0\| \gg \|vt + \mathfrak{R}_t r\|$, n can be approximated as $n = R_0 / \|R_0\|$, which is the direction of the radar line of sight (LOS). The Doppler frequency shift is then approximately

$$f_D = \frac{2f}{c}[v + \omega \times r] \cdot n$$

(2.64)

where the first term is the Doppler shift due to the translation

$$f_{\text{Trans}} = \frac{2f}{c}v \cdot n$$

(2.65)

and the second term is the micro-Doppler due to the rotation:

$$f_{mD} = \frac{2f}{c}[\omega \times r] \cdot n$$

(2.66)

For a time-varying rotation, the angular rotation velocity is a function of time and can be expressed by a polynomial function:

$$\Omega(t) = \Omega_0 + \Omega_1 t + \Omega_2 t^2 + \dots$$

(2.67)

If no more than second-order terms are used, the micro-Doppler shift can be expressed by

$$f_{mD} = \frac{2f}{c}[\Omega(t) \times r] \cdot n$$

$$= \frac{2f}{c}\left[\Omega_0 \cdot (r \times n) + \Omega_1 \cdot (r \times n)t + \Omega_2 \cdot (r \times n)t^2\right]$$

(2.68)

where the vector operation $(a \times b) \cdot c = a \cdot (b \times c)$ is applied.

2.4.2 Vibration-Induced Micro-Doppler Shift

As shown in Figure 2.10, the radar is located at the origin of the radar coordinate system (X, Y, Z) and a point scatterer P is vibrating about a center point

O. This center point is also the origin of the reference coordinate system (X', Y', Z'), which is translated from (X, Y, Z) to be situated at a distance R_0 from the radar. We also assume that the center point O is stationary with respect to the radar. If the azimuth and elevation angle of the point O with respect to the radar are α and β, respectively, the point O is located at

$$\left(R_0 \cos\beta\cos\alpha, R_0 \cos\beta\sin\alpha, R_0 \sin\beta\right) \tag{2.69}$$

in the radar coordinates (X, Y, Z). Then the unit vector of the radar LOS becomes

$$\boldsymbol{n} = \left[\cos\alpha\cos\beta, \sin\alpha\cos\beta, \sin\beta\right]^T \tag{2.70}$$

If one assumes that the scatterer P is vibrating at a frequency f_v with an amplitude D_v and that the azimuth and elevation angle of the vibration direction in the reference coordinates (X', Y', Z') are α_P and β_P, respectively. As shown in Figure 2.10, the vector from the radar to the scatterer P is $\boldsymbol{R}_t = \boldsymbol{R}_0 + \boldsymbol{D}_t$ and, thus, the range becomes

$$R_t = |\boldsymbol{R}_t| = \left[\left(R_0 \cos\beta\cos\alpha + D_t \cos\beta_P \cos\alpha_P\right)^2 \right.$$
$$+ \left(R_0 \cos\beta\sin\alpha + D_t \cos\beta_P \sin\alpha_P\right)^2 \tag{2.71}$$
$$\left. + \left(R_0 \sin\beta + D_t \sin\beta_P\right)^2 \right]^{1/2}$$

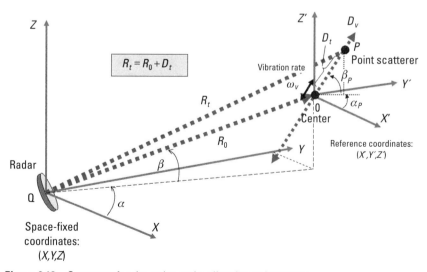

Figure 2.10   Geometry for the radar and a vibrating point target.

If $R_0 \gg D_t$, the range is approximately

$$R_t = \left\{ R_0^2 + D_t^2 + 2R_0 D_t \left[\cos\beta\cos\beta_p \cos\left(\alpha - \alpha_p\right) + \sin\beta\sin\beta_p \right] \right\}^{1/2}$$
$$\approx R_0 + D_t \left[\cos\beta\cos\beta_p \cos(\alpha - \alpha_p) + \sin\beta\sin\beta_p \right]$$

(2.72)

If the azimuth angle α of the center point O and the elevation angle β_p of the scatterer P are all zero, and if $R_0 \gg D_t$, then we have

$$R_t = \left(R_0^2 + D_t^2 + 2R_0 D_t \cos\beta\cos\alpha_p \right)^{1/2} \cong R_0 + D_t \cos\beta\cos\alpha_p \quad (2.73)$$

Because the angular frequency of the vibration rate is ω_v and the amplitude of the vibration is D_v, then $D_t = D_v \sin\omega_v t$ and the range of the scatterer becomes

$$R(t) = R_t = R_0 + D_v \sin\omega_v t \cos\beta\cos\alpha_p \quad (2.74)$$

The radar received signal then becomes

$$s_R(t) = \rho\exp\left\{ j\left[2\pi f t + 4\pi\frac{R(t)}{\lambda} \right] \right\} = \rho\exp\left\{ j\left[2\pi f t + \Phi(t) \right] \right\} \quad (2.75)$$

where ρ is the reflectivity of the point scatterer, f is the carrier frequency of the transmitted signal, λ is the wavelength, and $\Phi(t) = 4\pi R(t)/\lambda$ is the phase function.

Substituting (2.74) into (2.75) and denoting $B = (4\pi/\lambda)D_v\cos\beta\cos\alpha_p$, the received signal can be rewritten as

$$s_R(t) = \rho\exp\left\{ j\frac{4\pi}{\lambda}R_0 \right\}\exp\left\{ j2\pi f t + B\sin\omega_v t \right\} \quad (2.76)$$

Equation (2.76) can be further expressed in terms of the Bessel function of the first kind of order k:

$$J_k(B) = \frac{1}{2\pi}\int_{-\pi}^{\pi} \exp\left\{ j(B\sin u - ku)du \right\} \quad (2.77)$$

and thus

$$s_R(t) = \rho \exp\left(j\frac{4\pi}{\lambda} R_0 \right) \sum_{k=-\infty}^{\infty} J_k(B) \exp\left\{ j\left(2\pi ft + k\omega_v \right)t \right\}$$

$$= \rho \exp\left(j\frac{4\pi}{\lambda} R_0 \right) \Big\{ J_0(B)\exp(j2\pi ft) + J_1(B)\exp\left[j\left(2\pi f + \omega_v \right)t \right]$$

$$- J_1(B)\exp\left[j\left(2\pi f - \omega_v \right)t \right] + J_2(B)\exp\left[j\left(2\pi f + 2\omega_v \right)t \right]$$

$$- J_2(B)\exp\left[j\left(2\pi f - 2\omega_v \right)t \right] + J_3(B)\exp\left[j\left(2\pi f + 3\omega_v \right)t \right]$$

$$- J_3(B)\exp\left[j\left(2\pi f - 3\omega_v \right)t \right] + \ldots \Big\}$$

(2.78)

Therefore, the micro-Doppler frequency spectrum consists of pairs of spectral lines around the center frequency f and with a spacing $\omega_v/(2\pi)$ between adjacent lines.

Because of the vibration, the point scatterer P, which is initially, at time $t = 0$, located at $[X_0, Y_0, Z_0]^T$ in (X', Y', Z'), will, at time t, move to

$$\begin{bmatrix} X_1 \\ Y_1 \\ Z_1 \end{bmatrix} = D_v \sin(2\pi f_v t) \begin{bmatrix} \cos\alpha_P \cos\beta_P \\ \sin\alpha_P \cos\beta_P \\ \sin\beta_P \end{bmatrix} + \begin{bmatrix} X_0 \\ Y_0 \\ Z_0 \end{bmatrix} \qquad (2.79)$$

Due to the vibration, the velocity of the scatterer P becomes

$$v = 2\pi D_v f_v \cos(2\pi f_v t)\left[\cos\alpha_P \cos\beta_P, \sin\alpha_P \cos\beta_P, \sin\beta_P \right]^T \quad (2.80)$$

Based on the analysis in Section 2.4.1, the micro-Doppler shift induced by the vibration is

$$f_{mD}(t) = \frac{2f}{c}\left(v^T \cdot n \right)$$

$$= \frac{4\pi ff_v D_v}{c}\left[\cos(\alpha - \alpha_P)\cos\beta\cos\beta_P + \sin\beta\sin\beta_P \right]\cos(2\pi f_v t)$$

(2.81)

If the azimuth angle α and the elevation angle β_P are both zero, we have

$$f_{mD}(t) = \frac{4\pi ff_v D_v}{c}\cos\beta\cos\alpha_P \cos(2\pi f_v t) \qquad (2.82)$$

When the orientation of the vibrating scatterer is along the projection of the radar LOS direction, or $\alpha_p = 0$, and the elevation angle β is also 0, the Doppler frequency change reaches the maximum value of $4\pi f f_v D_v/c$.

2.4.3 Rotation-Induced Micro-Doppler Shift

The geometry of the radar and a target with 3-D rotations is illustrated in Figure 2.11. The radar coordinate system is (X, Y, Z), the target local coordinate system is (x, y, z), and the reference coordinate system (X', Y', Z') is parallel to the radar coordinates (X, Y, Z) and located at the origin of the target local coordinates. Assume that the azimuth and elevation angle of the target in the radar coordinates (X, Y, Z) are α and β, respectively, and the unit vector of the radar LOS is the same as (2.70).

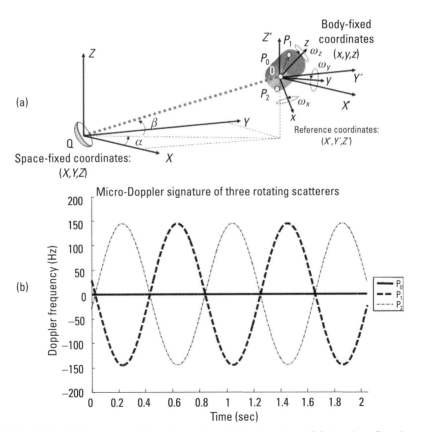

Figure 2.11 (a) Geometry of the radar and a rotating target, and (b) the micro-Doppler signature of the rotating target.

Due to the target's rotation, any point on the target described in the local coordinate system (x, y, z) will move to a new position in the reference coordinate system (X', Y', Z'). The new position can be calculated from its initial position vector multiplied by an initial rotation matrix of the x-convention (z-x-z sequence) determined by the angles $(\varphi_0, \theta_0, \psi_0)$, where the angle φ_0 rotates about the z-axis, the angle θ_0 rotates about the x-axis, and the angle ψ_0 rotates about the z-axis again.

The corresponding initial rotation matrix is defined by

$$\Re_{\text{Init}} = \Re_Z(\varphi_0) \cdot \Re_X(\theta_0) \cdot \Re_Z(\psi_0) = \begin{bmatrix} r_{11} & r_{12} & r_{13} \\ r_{21} & r_{22} & r_{23} \\ r_{31} & r_{32} & r_{33} \end{bmatrix} \qquad (2.83)$$

where

$$\begin{cases} r_{11} = -\sin\varphi_0 \cos\theta_0 \sin\psi_0 + \cos\varphi_0 \cos\psi_0 \\ r_{21} = -\cos\varphi_0 \cos\theta_0 \sin\psi_0 - \sin\varphi_0 \cos\psi_0 \\ r_{31} = \sin\theta_0 \sin\psi_0 \end{cases} \qquad (2.84)$$

$$\begin{cases} r_{12} = \sin\varphi_0 \cos\theta_0 \cos\psi_0 + \cos\varphi_0 \sin\psi_0 \\ r_{22} = \cos\varphi_0 \cos\theta_0 \cos\psi_0 - \sin\varphi_0 \sin\psi_0 \\ r_{32} = -\sin\theta_0 \cos\psi_0 \end{cases} \qquad (2.85)$$

$$\begin{cases} r_{13} = \sin\varphi_0 \sin\theta_0 \\ r_{23} = \cos\varphi_0 \sin\theta_0 \\ r_{33} = \cos\theta_0 \end{cases} \qquad (2.86)$$

Viewed in the target local coordinate system, when a target rotates about its axes x, y and z with an angular velocity $\boldsymbol{\omega} = (\omega_x, \omega_y, \omega_z)^T$, a point scatterer P at $\boldsymbol{r_P} = [x_P, y_P, z_P]^T$ represented in the target local coordinates (x, y, z) will move to a new location in the reference coordinate system described by $\Re_{\text{Init}} \cdot \boldsymbol{r_P}$ and the unit vector of the rotation becomes

$$\boldsymbol{\omega} = \left(\omega'_x, \omega'_y, \omega'_z \right)^T = \frac{\Re_{\text{Init}} \cdot \boldsymbol{\omega}}{\|\boldsymbol{\omega}\|} \qquad (2.87)$$

with the scalar angular velocity $\Omega = \|\boldsymbol{\omega}\|$. Thus, according to the Rodrigues formula [2], at time t the rotation matrix becomes

$$\Re_t = I + \hat{\omega}' \sin \Omega t + \hat{\omega}'^2 (1 - \cos \Omega t) \tag{2.88}$$

where $\hat{\omega}'$ is the skew symmetric matrix

$$\hat{\omega}' = \begin{bmatrix} 0 & -\omega'_z & \omega'_y \\ \omega'_z & 0 & -\omega'_x \\ -\omega'_y & \omega'_x & 0 \end{bmatrix} \tag{2.89}$$

Therefore, viewed in the reference coordinate system (X', Y', Z'), at time t, the scatterer P will move from its initial location to a new location $r = \Re_t \cdot \Re_{\text{Init}} \cdot r_P$. According to the discussion in Section 2.4.1, the micro-Doppler frequency shift induced by the rotation is approximately

$$f_{mD} = \frac{2f}{c} [\Omega \omega' \times r]_{\text{radial}} = \frac{2f}{c} (\Omega \hat{\omega}' r)^T \cdot n = \frac{2f}{c} [\Omega \hat{\omega}' \Re_t \cdot \Re_{\text{Init}} \cdot r_P]^T \cdot n$$

$$= \frac{2f\Omega}{c} \left\{ \left[\hat{\omega}'^2 \sin \Omega t - \hat{\omega}'^3 \cos \Omega t + \hat{\omega}' (I + \hat{\omega}'^2) \right] \Re_{\text{Init}} \cdot r_P \right\}^T \cdot n \tag{2.90}$$

If the skew symmetric matrix $\hat{\omega}'$ is defined by a unit vector, then $\hat{\omega}'^3 = -\hat{\omega}'$ and the rotation induced micro-Doppler frequency becomes

$$f_{mD} = \frac{2f\Omega}{c} [\hat{\omega}' (\hat{\omega}' \sin \Omega t + I \cos \Omega t) \Re_{\text{Init}} \cdot r_P]_{\text{radial}} \tag{2.91}$$

Assume that the radar operates at 10 GHz and a target, located at $(U = 1,000\text{m}, V = 5,000\text{m}, W = 5,000\text{m})$, is rotating along the x, y and z axes with the initial Euler angles $(\varphi = 30°, \theta = 20°, \psi = 20°)$ and angular velocity $\boldsymbol{\omega} = [\pi, 2\pi, \pi]^T$ rad/sec. Suppose that the target has three strong scatterer centers: scatterer P_0 (the center of the rotation) is located at $(x = 0\text{m}, y = 0\text{m}, z = 0\text{m})$; scatterer P_1 is located at $(x = 1.0\text{m}, y = 0.6\text{m}, z = 0.8\text{m})$; and scatterer P_2 is located at $(x = -1.0\text{m}, y = -0.6\text{m}, z = -0.8\text{m})$. The theoretical micro-Doppler modulation calculated by (2.91) is shown in Figure 2.11(b). The micro-Doppler of the center point P_0 is the line at the zero frequency, and the micro-Doppler modulations from the points P_1 and P_2 are the two sinusoidal curves about the zero frequency. The rotation period can be obtained from the rotation angular velocity as $T_0 = 2\pi / \|\boldsymbol{\omega}\| = 0.8165$ seconds.

2.4.4 Coning Motion-Induced Micro-Doppler Shift

A coning motion is a rotation about an axis that intersects with an axis of the local coordinates. A whipping top usually undergoes a coning motion while its body is spinning around its axis of symmetry with a fixed tip point and the axis of symmetry is rotating about another axis that intersects with the axis. If the axis of symmetry does not remain at a constant angle with the axis of coning, it will oscillate up and down between two limits, called nutation. In this case, the Euler angle φ is known as the spin angle, ψ is known as the precession angle, and θ is known as the nutation angle.

Without considering spinning and nutation, assume a target with pure coning motion along the axis SN, which intersects with the z-axis at the point $S(x = 0, y = 0, z = z_0)$ of the local coordinates, as shown in Figure 2.12. The reference coordinate system (X', Y', Z') is parallel to the radar coordinates (X, Y, Z) and its origin is located at the point S. Assume that the azimuth and elevation angle of the target center O with respect to the radar are α and β, respectively, and the azimuth and elevation angle of the coning axis SN with respect to the reference coordinates (X', Y', Z') are α_N and β_N, respectively. Then the unit vector of the radar LOS is

$$n = [\cos\alpha\cos\beta, \sin\alpha\cos\beta, \sin\beta]^T$$

and the unit vector of the rotation axis in the reference coordinates (X', Y', Z') is

$$e = \left[\cos\alpha_N \cos\beta_N, \sin\alpha_N \cos\beta_N, \sin\beta_N\right]^T$$

Assume the initial position of a scatterer P is $r = [x, y, z]^T$ represented in the target local coordinates. Then the location of the point scatterer P in the reference coordinates (X', Y', Z') can be calculated through its local coordinates by subtracting the coordinates of the point S and multiplying the rotation matrix $\mathfrak{R}_{\text{Init}}$ defined by the initial Euler angles (φ, θ, ψ). Viewed from the reference coordinates (X', Y', Z'), the location of the scatterer P is at $\mathfrak{R}_{\text{Init}} \cdot [x, y, z - z_0]^T$. Suppose that the target has a coning motion with an angular velocity of ω rad/sec. According to the Rodrigues formula, at time t the rotation matrix in the reference coordinates (X', Y', Z') becomes

$$\mathfrak{R}_t = I + \hat{e}\sin\omega t + \hat{e}^2(1 - \cos\omega t) \tag{2.92}$$

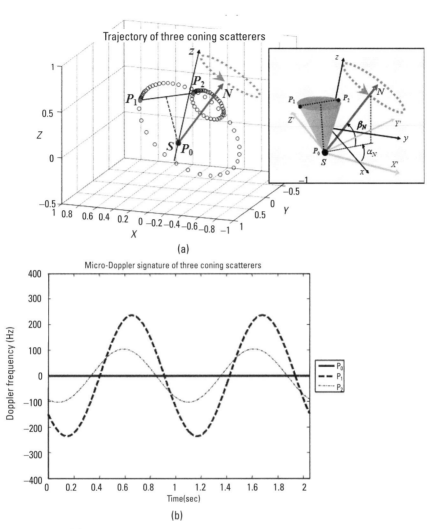

Figure 2.12 (a) A target with pure coning motion along an axis that intersects with the z-axis at the origin of the local coordinates, and (b) the micro-Doppler signature of the coning target.

where the skew symmetric matrix is defined by

$$
\hat{e} = \begin{bmatrix}
0 & -\sin\beta_N & \sin\alpha_N\cos\beta_N \\
\sin\beta_N & 0 & -\cos\alpha_N\cos\beta_N \\
-\sin\alpha_N\cos\beta_N & \cos\alpha_N\cos\beta_N & 0
\end{bmatrix} \quad (2.93)
$$

Therefore, at time t the scatterer P will move to

$$r(t) = \Re_t \cdot \Re_{\text{Init}} \cdot \left[x, y, z - z_0 \right]^T \tag{2.94}$$

If the point S is not too far from the target center of mass, the radar LOS can be approximated as the radial direction of the point S with respect to the radar. According to the mathematical formula described in Section 2.4.1, the micro-Doppler modulation induced by the coning motion is approximately

$$f_{mD} = \frac{2f}{c} \left[\frac{d}{dt} r(t) \right]_{\text{radial}} = \frac{2f}{c} \left\{ \left[\frac{d}{dt} \Re_t \right] \Re_{\text{Init}} \cdot \left[x, y, z - z_0 \right]^T \right\}^T \cdot n$$

$$= \frac{2f\omega}{c} \left\{ \left(\hat{e} \cos \omega t + \hat{e}^2 \sin \omega t \right) \Re_{\text{Init}} \cdot \left[x, y, z - z_0 \right]^T \right\}^T \cdot n \tag{2.95}$$

Assume that the radar operates at 10 GHz and a target is initially located at ($X = 1{,}000$m, $Y = 5{,}000$m, $Z = 5{,}000$m). Suppose that the initial Euler angles are $\varphi = 10°$, $\theta = 10°$ and $\psi = -20°$, the target is coning with 2 Hz during $T = 2.048$ seconds, or with an angular velocity $\omega = (2 \times 2\pi/T)$ rad/sec, and the azimuth and elevation angle of the rotation axis are $\alpha_N = 160°$ and $\beta_N = 50°$, respectively. Thus, given the initial location of the point scatterer P_0 at ($x = 0$m, $y = 0$m, $z = 0$ m), the scatterer P_1 at ($x = 0.3$m, $y = 0$m, $z = 0.6$m), and the scatterer P_2 at ($x = -0.3$m, $y = 0$m, $z = 0.6$m), the theoretical micro-Doppler modulation calculated by (2.95) is shown in Figure 2.12(b). The micro-Doppler of the point P_0 is zero frequency, and the micro-Doppler modulations from the points P_1 and P_2 are the two sinusoidal curves about the zero frequency with their amplitudes determined by their radial directions.

When a coning body is also spinning around its axis of symmetry, the motion becomes precession. The precession matrix $\Re_{\text{prec}}$ is the product of the coning matrix $\Re_{\text{coning}}$ and the spinning matrix $\Re_{\text{spinning}}$:

$$\Re_{\text{prec}} = \Re_{\text{coning}} \cdot \Re_{\text{spinning}} \tag{2.96}$$

where the coning matrix is

$$\Re_{\text{coning}} = I + \hat{\omega}_{\text{coning}} \sin\left(\Omega_{\text{coning}} t \right) + \hat{\omega}^2_{\text{coning}} \left[1 - \cos\left(\Omega_{\text{coning}} t \right) \right] \tag{2.97}$$

and the spinning matrix is

$$\Re_{\text{spinning}} = I + \hat{\omega}_{\text{spinning}} \sin\left(\Omega_{\text{spinning}} t \right) + \hat{\omega}^2_{\text{spinning}} \left[1 - \cos\left(\Omega_{\text{spinning}} t \right) \right] \tag{2.98}$$

The Ω_{coning} and Ω_{spinning} are the coning and spinning angular velocity, respectively. The skew symmetric matrix of the coning motion is defined by

$$
\hat{\omega}_{\text{coning}} = \begin{bmatrix} 0 & -\omega_{\text{coning}z} & \omega_{\text{coning}y} \\ \omega_{\text{coning}z} & 0 & -\omega_{\text{coning}x} \\ -\omega_{\text{coning}y} & \omega_{\text{coning}x} & 0 \end{bmatrix} \tag{2.99}
$$

and the skew symmetric matrix of the spinning motion is defined by

$$
\hat{\omega}_{\text{spinning}} = \begin{bmatrix} 0 & -\omega_{\text{spinning}z} & \omega_{\text{spinning}y} \\ \omega_{\text{spinning}z} & 0 & -\omega_{\text{spinning}x} \\ -\omega_{\text{spinning}y} & \omega_{\text{spinning}x} & 0 \end{bmatrix} \tag{2.100}
$$

A given point scatterer P at $r_P = [x_P, y_P, z_P]^T$ represented in the target local coordinate system (x, y, z) will move to a new location in the reference coordinate system described by $\Re_{\text{Init}} \cdot r_P$. Viewed in the reference coordinate system (X', Y', Z'), at time t, the scatterer P will move from its initial location to a new location $r = \Re_t \cdot \Re_{\text{Init}} \cdot r_P$.

Thus, due to precession, the micro-Doppler modulation of the point scatterer P is formulated as the following:

$$
\begin{aligned}
f_{mD}|_P &= \frac{2f}{c} \left[\frac{d}{dt} r \right]_{\text{radial}} = \frac{2f}{c} \left[\frac{d}{dt} \left(\Re_{\text{prec}} \cdot \Re_{\text{Init}} \cdot r_P \right) \right]_{\text{radial}} \\
&= \frac{2f}{c} \left[\frac{d}{dt} \left(\Re_{\text{coning}} \cdot \Re_{\text{spinning}} \right) \cdot \Re_{\text{Init}} \cdot r_P \right]_{\text{radial}} \\
&= \frac{2f}{c} \left[\left(\frac{d}{dt} \Re_{\text{coning}} \cdot \Re_{\text{spinning}} + \Re_{\text{coning}} \cdot \frac{d}{dt} \Re_{\text{spinning}} \right) \cdot \Re_{\text{Init}} \cdot r_P \right]_{\text{radial}}
\end{aligned} \tag{2.101}
$$

Assume that the radar operates at 10 GHz and a target is initially located at $(X = 1{,}000\text{m}, Y = 5{,}000\text{m}, Z = 5{,}000\text{m})$. Suppose that the initial Euler angles are $\varphi = 30°$, $\theta = 30°$ and $\psi = 20°$, and the target is spinning with 8 Hz during $T = 2.048$ seconds, or with an angular velocity $\Omega_{\text{spinning}} = (8 \times 2\pi/T)$ rad/sec. Then, given the initial location of the point scatterer P_0 at $(x = 0\text{m}, y = 0\text{m}, z = 0\text{m})$, the scatterer P_1 at $(x = 0.3\text{m}, y = 0\text{m}, z = 0.6\text{m})$, and the scatterer P_2 at $(x = -0.3\text{m}, y = 0\text{m}, z = 0.6\text{m})$, the theoretical micro-Doppler signature of the spinning target is shown in Figure 2.13.

Trajectory of a spinning target with three scatterers

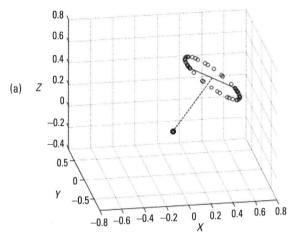

(a)

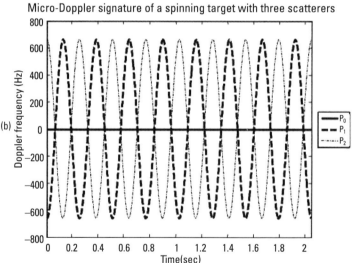

(b)

Figure 2.13   (a) A target with spinning motion, and (b) the micro-Doppler signature of the spinning target.

Based on (2.101), the theoretical micro-Doppler signature of a precession target can be seen in Figure 2.14, where the target has the same initial condition as the coning and the spinning target as discussed above. However, the target is spinning with an angular velocity $\Omega_{spinning} = 3.906$ rad/sec or spinning 8 cycles during $T = 2.048$ seconds. The target is also coning with an angular velocity $\Omega_{coning} = 0.977$ rad/sec or coning 2 cycles during $T = 2.048$ seconds.

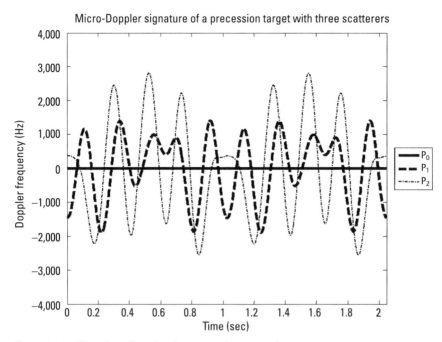

Figure 2.14 The micro-Doppler signature of a precession target.

The precession micro-Doppler signature shows 8 cycles spinning modulated by 2 cycles coning.

2.5 Bistatic Micro-Doppler Effect

The micro-Doppler effect is dependent on the target aspect angle. For a target moving at angles 0° (head-on) and 180° (tail-on) to the radar LOS, micro-Doppler frequencies reach the maximum shift; and at angles ±90° to the radar LOS, micro-Doppler frequencies become zero. Besides, due to occlusions at certain aspect angles, some corresponding parts of the target may not be seen by radar.

A separated transmitter and receiver bistatic configuration may acquire more targets' information complementary to that by using a monostatic radar, avoid blind velocities, and prevent null or low monostatic RCS positions.

In bistatic radar systems, to determine the target location, the azimuth and elevation angle of the target viewed at the transmitter and the timing of the transmitted signal must be known exactly. The synchronization between the receiver and transmitter must be maintained. In a bistatic radar, the range resolution and Doppler resolution are both dependent on the bistatic angle (i.e., the angle between the transmitter-to-target line and the receiver-to-target line). Rotational motion of the target can also induce a bistatic micro-Doppler effect, which depends on the triangulation geometry of the transmitter, the target, and the receiver.

In a bistatic radar, the transmitter and receiver are separated by a baseline distance that is comparable with the maximum range of the target with respect to the transmitter and the receiver. The separation brings an issue of synchronization between the two sites and requires phase synchronization between the local oscillator in the transmitter and the one in the receiver to accurately measure the target location and to perform range and Doppler processing.

The coordinate systems include global space-fixed coordinates, reference coordinates, and target local body-fixed coordinates, as shown in Figure 2.15 in a 3-D case. The bistatic plane is the one on which the transmitter, the receiver, and the target lie. The baseline (L) is the distance between the transmitter and receiver. The range from the transmitter to the target is a vector $\mathbf{r}_T$ and the range from the receiver to the target is a vector $\mathbf{r}_R$. The bistatic angle β is the angle between the transmitter-to-target line and the receiver-to-target line. Assume that the transmitter look angles (azimuth and elevation) are (α_T, β_T), the receiver look angles (α_R, β_R) can be obtained from the baseline distance, the target range, and the look angles relative to the transmitter. The positive angle is defined in a counterclockwise direction. Thus, the distance from the receiver to the target becomes

$$r_R = |\mathbf{r}_R| = \left(L^2 + r_T^2 \cos^2 \beta_T - 2r_T L \cos \beta_T \sin \alpha_T\right)^{1/2} \qquad (2.102)$$

and the receiver look angles are

$$\alpha_R = \tan^{-1}\left(\frac{L - r_T \cos \beta_T \sin \alpha_T}{r_T \cos \beta_T \cos \alpha_T}\right) \qquad (2.103)$$

$$\beta_R = \tan^{-1}\left[\frac{r_T \sin \beta_T}{\left(L^2 + r_T^2 \cos^2 \beta_T - 2r_T L \cos \beta_T \sin \alpha_T\right)^{1/2}}\right] \qquad (2.104)$$

The received signal from a point scatterer P can be modeled as

$$s_r^P(t) = \rho(\mathbf{r}_P)\exp\left\{j2\pi f \frac{|\mathbf{r}_T(t) + \mathbf{r}_P(t)| + |\mathbf{r}_R(t) + \mathbf{r}_P(t)|}{c}\right\} \qquad (2.105)$$

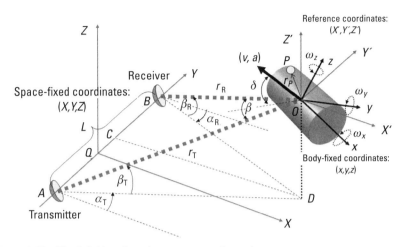

Figure 2.15 The 3-D bistatic radar system configuration.

where c is the wave propagation speed. Then the received signal from a volume target is a volume integration of the point return over the whole target:

$$s_r(t) = \iiint\limits_{\text{Target}} \rho(\mathbf{r}_P) \exp\left\{ j2\pi f \frac{|\mathbf{r}_T(t) + \mathbf{r}_P(t)| + |\mathbf{r}_R(t) + \mathbf{r}_P(t)|}{c} \right\} d\mathbf{r}_P \qquad (2.106)$$

The phase term in the received signal is

$$\begin{aligned}
\Phi_P(t) &= 2\pi f \frac{|\mathbf{r}_T(t) + \mathbf{r}_P(t)| + |\mathbf{r}_R(t) + \mathbf{r}_P(t)|}{c} \\
&= 2\pi f \frac{r_T(t) + r_R(t) + \left[\mathbf{r}_T(t) + \mathbf{r}_R(t)\right] \cdot \mathbf{r}_P(t)}{c} \\
&= 2\pi f \frac{r_T(t) + r_R(t)}{c} \cdot 2\pi f \frac{\left[\mathbf{r}_T(t) + \mathbf{r}_R(t)\right] \cdot \mathbf{r}_P(t)}{c} \\
&= \Phi_V(t) \cdot \Phi_{\Omega,P}(t)
\end{aligned} \qquad (2.107)$$

where

$$\Phi_V(t) = 2\pi f \frac{r_T(t) + r_R(t)}{c} \qquad (2.108)$$

is the phase term induced by the translational motion and

$$\Phi_{\Omega,P}(t) = 2\pi f \frac{\left[r_T(t) + r_R(t)\right] \cdot r_P(t)}{c} \tag{2.109}$$

is the phase term induced by the rotational motion of the point scatterer P.

While the target has translational motion with a velocity V and rotation with an initial Euler angle $(\varphi_0, \theta_0, \psi_0)$ and an angular velocity vector $\Omega = (\omega_x, \omega_y, \omega_z)^T$ rotating about the target local-fixed axes x, y, and z, then the Doppler frequency shift of the point scatterer P induced by the translation and rotation can be obtained by the time derivative of the phase function.

The Doppler shift consists of two parts: one is induced by the translation and the other is induced by the rotation,

$$f_{D_{Bi}} = f_{D_{Trans}} + f_{D_{Rot}} \tag{2.110}$$

where

$$f_{D_{Trans}} = \frac{f}{c} \frac{d}{dt}\left[r_T(t) + r_R(t)\right] \tag{2.111}$$

is the translational Doppler shift and

$$f_{D_{Rot}} = \frac{2f}{c}\left[\Omega \times r_P(t)\right] \tag{2.112}$$

is the rotation induced micro-Doppler shift, which is usually a periodic frequency function of time and distributed around the translational Doppler frequency.

If the target moves with a velocity V and an acceleration a, their components along the direction from the transmitter to the target are

$$V_T = V \cdot \frac{r_T}{|r_T|}; \quad a_T = a \cdot \frac{r_T}{|r_T|} \tag{2.113}$$

and the components along the direction from the receiver to the target are

$$V_R = V \cdot \frac{r_R}{|r_R|}; \quad a_R = a \cdot \frac{r_R}{|r_R|} \tag{2.114}$$

Then the range from the transmitter to the moving target becomes

$$r_T = r_{T0} + V_T t + \frac{1}{2} a_T t^2 + \dots \tag{2.115}$$

and the range from the receiver to the target is

$$r_R = r_{R0} + V_R t + \frac{1}{2} a_R t^2 + \dots \tag{2.116}$$

While the target has rotational motion, its rotation angle is determined by its initial angle $\boldsymbol{\theta}_0$ and rotation rate $\boldsymbol{\Omega}$:

$$\boldsymbol{\theta} = \boldsymbol{\theta}_0 + \boldsymbol{\Omega} t + \dots \tag{2.117}$$

where

$$\boldsymbol{\Omega} = \left(\omega_x, \omega_y, \omega_z \right) \tag{2.118}$$

Therefore, the Doppler shift induced by the translational motion is

$$f_{D_{\text{Tran}}} = \frac{f}{c} \frac{d}{dt} \left[r_T(t) + r_R(t) \right] = \frac{f}{c} \left[V_T + V_R + \left(a_T + a_R \right) t \right] \tag{2.119}$$

and that induced by the rotation becomes

$$f_{D_{\text{Rot}}} = \frac{2\pi f}{c} \frac{d}{dt} \left\{ \left[\mathbf{r}_T(t) + \mathbf{r}_R(t) \right] \cdot \mathbf{r}_P(t) \right\} \tag{2.120}$$

Thus, the translational Doppler shift of the bistatic radar depends on three factors [22, 23]. The first factor is the maximum Doppler shift when a target moving with a velocity V:

$$f_{D_{\text{Max}}} = \frac{2f}{c} |V| \tag{2.121}$$

The second factor is related to the bistatic triangulation factor

$$D = \cos\left(\frac{\alpha_R - \alpha_T}{2} \right) = \cos\left(\frac{\beta}{2} \right) \tag{2.122}$$

where $\beta = \alpha_R - \alpha_T$ is the bistatic angle.

The third factor is related to angle δ between the moving direction of the target and the direction of the bisector:

$$C = \cos\delta \qquad (2.123)$$

Thus, the bistatic Doppler shift becomes

$$f_{D_{Bi}} = f_{D_{Max}} D \cdot C = \frac{2f}{c}|V|\cos\left(\frac{\beta}{2}\right)\cos\delta \qquad (2.124)$$

Similar to the Doppler shift induced by the radial velocity in the monostatic radar case, for the bistatic radar case the Doppler shift is induced by the target's bisector velocity. The bistatic Doppler shift is always smaller than the maximum monostatic Doppler shift because the term, $\cos(\beta/2)$, is always less than 1.

If the radial velocity of the target with respect to the transmitter is very small or zero, the monostatic radar cannot measure the radial velocity of the target. However, according to (2.124), the target velocity with respect to the receiver may not be zero. Thus, the bistatic radar is able to measure the bisector velocity of the moving target.

As defined in the monostatic radar, in the bistatic case two targets may be separated in range, Doppler shift, and angle. The range resolution in the monostatic radar is Δr_{Mono} and the Doppler resolution is Δf_{DMono}. For bistatic radar, the range resolution and Doppler resolution are both dependent on the bistatic angle β. The range resolution becomes

$$\Delta r_{Bi} = \frac{1}{\cos(\beta/2)}\Delta r_{Mono} \qquad (2.125)$$

and the Doppler resolution is

$$\Delta f_{D_{Bi}} = \cos\left(\frac{\beta}{2}\right)\Delta f_{D_{Mono}} \qquad (2.126)$$

If the bistatic angle is near 180°, this is the case of forward scattering radar. Cherniakov discussed the advantages and disadvantages of the forward scattering radar [24]. For bistatic angles over 150°–160°, the bistatic RCS increases sharply. However, for the bistatic angle near 180°, $\cos(\beta/2) \to 0$. Thus, the bistatic range resolution is losing and there is no Doppler frequency resolution.

Similar to the bistatic Doppler shift generated by the bisector velocity, for a point scatterer P in a target, its bistatic micro-Doppler induced by the rotational motion of the target is determined by its bisector component [22]:

$$f_{mD_{Bi}}(t,P) = f_{D_{Rot}}(t,P) = \frac{2f}{c}\left[\Omega \times r_p(t)\right]_{Bisector} \tag{2.127}$$

The initial Euler angle $(\varphi_0, \theta_0, \psi_0)$, the angular velocity vector $\Omega = (\omega_x, \omega_y, \omega_z)^T$, and the point scatterer's location P determine the total micro-Doppler shift.

According to (2.127), the bistatic micro-Doppler shift of a rotating target can be expressed by

$$f_{mD_{Bi}} = f_{mD_{Max}} \cos\left(\frac{\beta}{2}\right) \tag{2.128}$$

where

$$f_{mD_{Max}} = \frac{2f}{c}\left\|\Omega \times r_p(t)\right\| \tag{2.129}$$

is the maximum micro-Doppler captured by the monostatic radar.

2.6 Multistatic Micro-Doppler Effect

The multistatic radar configuration has multiple transmitters and receivers distributed over several locations called nodes. It combines multiple measurements of a target viewed at different aspects to extract information about the target. Any individual node of the multistatic system may have both a transmitter and a receiver [24].

In a multistatic radar system, the range of a target is determined by a pair of a transmitter and a receiver and is measured from the time delay Δt between the transmitted and the received signal $r_{bistatic} = r_T + r_R = c \cdot \Delta t$. However, the location of the target measured by the bistatic radar is ambiguous and constrained to an ellipse with foci at the transmitter and the receiver.

The multistatic system can be treated as combinations of several bistatic systems. Figure 2.16 illustrates a multistatic system with four transmitters and four receivers. There is a total of $N_{Channel} = N_{Trans} \times N_{Receiv} = 16$ possible channels to form the multistatic system: four monostatic systems and twelve

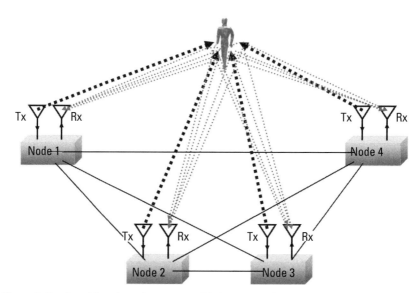

Figure 2.16 A multistatic radar system with four transmitters and four receivers.

bistatic systems. Because each bistatic system provides a different aspect view of the target, the multistatic system will provide a simultaneous multiple view of the micro-Doppler effect [25].

The processing in the multistatic system is to coherently combine the signals received from each of these bistatic systems, model the change of the target position with time, and calculate the phase change $\Phi_n(t)$ in each bistatic system. Then the received baseband signal can be expressed by

$$s_r(t) = \sum_{n=1}^{N} A_n \exp\{-j\Phi_n(t)\} = \sum_{n=1}^{N} A_n \exp\left\{-j\frac{2\pi f\left[r_{T,n}(t) + r_{R,n}(t)\right]}{c}\right\} \quad (2.130)$$

where A_n is the amplitude and $\Phi_n(t)$ is the phase function of the signal in the nth channel, $r_{T,n}(t) = \|\mathbf{r}_{T,n}(t)\|$ is the distance from the nth transmitter to the target, and $r_{R,n}(t) = \|\mathbf{r}_{R,n}(t)\|$ is the distance from the target to the nth receiver.

If a point P on the target has rotation, the micro-Doppler induced by the rotation is given by

$$f_{mD_{Bi}}(t,P) = f_{D_{Rot}}(t,P) = \frac{2f}{c}\left[\mathbf{\Omega} \times \mathbf{r}_P(t)\right]_{Bisector} \quad (2.131)$$

as derived in (2.127).

Thus, the multistatic micro-Doppler of the point P is derived by taking the time derivative of the combined baseband signals

$$f_{mD_{\text{Multi}}}(t,P) = \sum_{n=1}^{N} A_n f_{mD_{\text{Bi}}}(t,P) \qquad (2.132)$$

In the multistatic configuration, each node has its own aspect view of the target. The multistatic topology determines the multistatic micro-Doppler features. The combination of the number of nodes and the angular separation between these nodes determine the total micro-Doppler signature of the point P.

2.7   Cramer-Rao Bound of the Micro-Doppler Estimation

In a coherent radar system, the micro-Doppler modulation is embedded in the radar returns. The range profile can be expressed by

$$r(t) = R_0 + d \cdot \cos\omega_{mD} t \qquad (2.133)$$

where R_0 is the distance of the target, d is the displacement or the amplitude of the micro-Doppler modulation, and ω_{mD} is the micro-Doppler modulation frequency. To estimate micro-Doppler frequency shifts, the signal-to-noise ratio (*SNR*), the total number of time samples (*N*), the frequency of micro-Doppler modulation (ω_{mD}), the displacement of the micro-Doppler modulation (d), and the wavelength of the carrier frequency (λ) determine the lower bound on the micro-Doppler estimation. The Cramer-Rao lower bound of the micro-Doppler estimation can be found in [26, 27].

References

[1]   Goldstein, H., *Classical Mechanics*, 2nd ed., Reading, MA: Addison-Wesley, 1980.

[2]   Murray, R. M., Z. Li, and S. S. Sastry, *A Mathematical Introduction to Robotic Manipulation*, Boca Raton, FL: CRC Press, 1994.

[3]   Wittenburg, J., *Dynamics of Systems of Rigid Bodies*, Stuttgart: Teubner, 1977.

[4]   Kuipers, J. B., *Quaternions and Rotation Sequences*, Princeton, NJ: Princeton University Press, 1999.

[5]   Mukundand, R., "Quaternions: From Classical Mechanics to Computer Graphics, and Beyond," *Proc. of 7th Asian Technology Conference in Mathematics (ATCM)*, 2002, pp. 97–106.

[6]   Shoemake, K., "Animating Rotation with Quaternion Curves," *ACM Computer Graphics*, Vol. 19, No. 3, 1985, pp. 245–254.

[7]   Klumpp, A. R., "Singularity-Free Extraction of a Quaternion from a Direction-Cosine Matrix," *Journal of Spacecraft and Rockets*, Vol. 13, 1976, pp. 754–755.

[8]   Géradin, M., and A. Cardona, *Flexible Multibody Dynamics: A Finite Element Approach*, New York: John Wiley & Sons, 2001.

[9]   Shabana, A. A., *Dynamics of Multibody Systems*, 3rd ed., Cambridge, U.K.: Cambridge University Press, 2005.

[10]   Magnus, K., *Dynamics of Multibody Systems*, New York: Springer-Verlag, 1978.

[11]   Knott, E. F., J. F. Schaffer, and M. T. Tuley, *Radar Cross Section*, 2nd ed., Norwood, MA: Artech House, 1993.

[12]   Ruck, G. T., et al., *Radar Cross Section Handbook*, New York: Plenum Press, 1970.

[13]   Jenn, D., "Radar Cross-Section," in *Encyclopedia of RF and Microwave Engineering*, K. Chang, (ed.), New York: John Wiley & Sons, 2005.

[14]   Ling, H., K. Chou, and S. Lee, "Shooting and Bouncing Rays: Calculating the RCS of an Arbitrarily Shaped Cavity," *IEEE Transactions on Antennas and Propagation*, Vol. 37, No. 2, 1989, pp. 194–205.

[15]   Kunz, K., and R. Luebbers, *The Finite-Difference Time Domain Method for Electromagnetics*, Boca Raton, FL: CRC Press, 1993.

[16]   Chatzigeorgiadis, F., and D. Jenn, "A MATLAB Physical-Optics RCS Prediction Code," *IEEE Antennas and Propagation Magazine*, Vol. 46, No. 4, 2004, pp. 137–139.

[17]   Chatzigeorgiadis, F., "Development of Code for Physical Optics Radar Cross Section Prediction and Analysis Application," Master's Thesis, Naval Postgraduate School, Monterey, CA, September 2004.

[18]   Cooper, J., "Scattering by Moving Bodies: The Quasi-Stationary Approximation," *Mathematical Methods in the Applied Sciences*, Vol. 2, No. 2, 1980, pp. 131–148.

[19]   Kleinman, R. E., and R. B. Mack, "Scattering by Linearly Vibrating Objects," *IEEE Transactions on Antennas and Propagation*, Vol. 27, No. 3, 1979, pp. 344–352.

[20]   Van Bladel, J., "Electromagnetic Fields in the Presence of Rotating Bodies," *Proc. of the IEEE*, Vol. 64, No. 3, 1976, pp. 301–318.

[21]   Chen, V. C., et al., "Micro-Doppler Effect in Radar: Phenomenon, Model, and Simulation Study," *IEEE Transactions on Aerospace and Electronics Systems*, Vol. 42, No. 1, 2006, pp. 2–21.

[22]   Chen, V. C., A. des Rosiers, and R. Lipps, "Bi-Static ISAR Range-Doppler Imaging and Resolution Analysis," *IEEE Radar Conference*, Pasadena, CA, May 2009.

[23]   Willis, N. J., *Bistatic Radar*, 2nd ed., Raleigh, NC: SciTech Publishing, 2005.

[24] Chernyak, V. S., *Fundamentals of Multisite Radar Systems: Multistatic Radars and Multiradar Systems*, Amsterdam, the Netherlands: Gordon and Breach Scientific Publishers, 1998.

[25] Chen, V. C., et al., "Radar Micro-Doppler Signatures for Characterization of Human Motion," Chapter 15 in *Through-the-Wall Radar Imaging*, M. Amin, (ed.), Boca Raton, FL: CRC Press, 2010.

[26] Rao, C. R., "Information and Accuracy Attainable in the Estimation of Statistical Parameters," *Bull. Calcutta Math. Soc.*, No. 37, 1945, pp. 81–91.

[27] Setlur, P., M. Amin, and F. Ahmad, "Optimal and Suboptimal Micro-Doppler Estimation Schemes Using Carrier Diverse Doppler Radars," *Proc. of the IEEE International Conference on Acoustics, Speech and Signal Processing*, Taipei, Taiwan, 2009.

Appendix 2A

For any vector $\boldsymbol{u} = [u_x, u_y, u_z]^T$, the skew symmetric matrix is defined by

$$\hat{u} = \begin{bmatrix} 0 & -u_z & u_y \\ u_z & 0 & -u_x \\ -u_y & u_x & 0 \end{bmatrix} \tag{2A.1}$$

The cross-product of two vectors $\boldsymbol{u}$ and $\boldsymbol{r}$ can be derived through matrix computation as

$$\boldsymbol{p} = \boldsymbol{u} \times \boldsymbol{r} = \begin{bmatrix} u_y r_z - u_z r_y \\ u_z r_x - u_x r_z \\ u_x r_y - u_y r_x \end{bmatrix} = \begin{bmatrix} 0 & -u_z & u_y \\ u_z & 0 & -u_x \\ -u_y & u_x & 0 \end{bmatrix} \cdot \begin{bmatrix} r_x \\ r_y \\ r_z \end{bmatrix} = \hat{u} \cdot \boldsymbol{r} \tag{2A.2}$$

The cross-product's definition is useful in analyzing special orthogonal matrix groups called the $SO(3)$ groups or the 3-D rotation matrix, defined by:

$$SO(3) = \left\{ R \in \Re^{3\times3} \,|\, R^T R = I, \det(R) = +1 \right\} \tag{2A.3}$$

Computing the derivative of the constraint $R(t)R^T(t) = I$ with respect to time t, a differential equation can be obtained as

$$\dot{R}(t)R^T(t) + R(t)\dot{R}^T(t) = 0 \tag{2A.4}$$

or

$$\dot{R}(t)R^T(t) = -\left[\dot{R}(t)R^T(t) \right]^T \tag{2A.5}$$

The result reflects the fact that the matrix $\dot{R}(t)R^T(t) \in \Re^{3\times3}$ is a skew symmetric matrix. Thus, there must be a vector $\boldsymbol{\omega} \in \Re^3$ such that:

$$\hat{\omega} = \dot{R}(t)R^T(t) \tag{2A.6}$$

Multiplying both sides by $R(t)$ to the right yields

$$\dot{R}(t) = \hat{\omega}R(t) \tag{2A.7}$$

Assume that the vector $\boldsymbol{\omega} \in \mathfrak{R}^3$ is constant. According to the linear ordinary differential equation (ODE), the solution becomes

$$R(t) = \exp\{\hat{\omega}t\}R(0) \tag{2A.8}$$

where $\exp\{\hat{\omega}t\}$ is the matrix exponential:

$$\exp\{\hat{\omega}t\} = I + \hat{\omega}t + \frac{(\hat{\omega}t)^2}{2!} + \cdots + \frac{(\hat{\omega}t)^n}{n!} + \cdots$$

Assuming for the initial condition: $R(0) = I$, (2A.8) becomes

$$R(t) = \exp\{\hat{\omega}t\} \tag{2A.9}$$

Thus, it can be conformed that the matrix $\exp\{\hat{\omega}t\}$ is indeed a rotation matrix. Since

$$\left[\exp(\omega t)\right]^{-1} = \exp(-\hat{\omega}t) = \exp\left(\hat{\omega}^T t\right) = \left[\exp(-\hat{\omega}t)\right]^T \tag{2A.10}$$

and thus $[\exp(\hat{\omega}t)]^T \cdot \exp(\hat{\omega}t) = I$, from which one can obtain $\det\{\exp(\hat{\omega}t)\} = \pm 1$.

Furthermore,

$$\det\left\{\exp(\hat{\omega}t)\right\} = \det\left\{\exp\left(\frac{\hat{\omega}t}{2}\right) \cdot \exp\left(\frac{\hat{\omega}t}{2}\right)\right\} = \left[\det\left\{\exp\left(\frac{\hat{\omega}t}{2}\right)\right\}\right]^2 \geq 0 \tag{2A.11}$$

which shows that $\det\{\exp(\hat{\omega}t)\} = \pm 1$. Therefore, matrix $R(t) = \exp\{\hat{\omega}t\}$ is the 3-D rotation matrix. Let $\Omega = \|\boldsymbol{\omega}\|$. A physical interpretation of the equation: $R(t) = \exp\{\hat{\omega}t\}$ is simply a rotation around the axis $\boldsymbol{\omega} \in \mathfrak{R}^3$ by $\Omega \cdot t$ radians. If the rotation axis and the scalar angular velocity are given by a vector $\boldsymbol{\omega} \in \mathfrak{R}^3$, the rotation matrix can be computed as $R(t) = \exp\{\hat{\omega}t\}$ at time t.

The Rodrigues's formula is one efficient way to compute the rotation matrix $R(t) = \exp\{\hat{\omega}t\}$. Given $\boldsymbol{\omega}' \in \mathfrak{R}^3$ with $\|\boldsymbol{\omega}'\| = 1$ and $\boldsymbol{\omega} = \Omega \cdot \boldsymbol{\omega}'$, it is simple to verify that the power of $\hat{\omega}'$ can be reduced by the following formula:

$$\hat{\omega}'^3 = -\hat{\omega}' \tag{2A.12}$$

Then the exponential series

$$\exp(\hat{\omega}t) = I + \hat{\omega}t + \frac{(\hat{\omega}t)^2}{2!} + \cdots + \frac{(\hat{\omega}t)^n}{n!} + \cdots \qquad (2A.13)$$

can be simplified as:

$$\exp(\hat{\omega}t) = I + \left(\Omega t - \frac{(\Omega t)^3}{3!} + \frac{(\Omega t)^5}{5!} - \cdots\right)\hat{\omega}' + \left(\frac{(\Omega t)^2}{2!} - \frac{(\Omega t)^4}{4!} + \frac{(\Omega t)^6}{6!} - \cdots\right)\hat{\omega}'^2$$

$$= I + \hat{\omega}'\sin\Omega t + \hat{\omega}'^2(1 - \cos\Omega t)$$

$$(2A.14)$$

Therefore,

$$R(t) = \exp(\hat{\omega}t) = I + \hat{\omega}'\sin\Omega t + \hat{\omega}'^2(1 - \cos\Omega t) \qquad (2A.15)$$

3

The Micro-Doppler Effect of Rigid Body Motion

The rigid body is an idealization of a solid body without deformation (i.e., the distance between any two particles of the body remains constant while in motion). Usually, the geometry of a body is described by its location and orientation. The location is defined by the position of a reference point in the body, such as the center of mass or the centroid. The orientation is determined by its angular position. Thus, the motion of a rigid body is described by its kinematic and dynamic quantities, such as linear and angular velocity, linear and angular acceleration, linear and angular momentum, and the kinetic energy of the body. As described in Chapter 2, the orientation of a rigid body can be represented by a set of Euler angles in a three-dimensional (3-D) Euclidean space, by a rotation matrix (called the direction cosine matrix), or by a quaternion.

When a rigid body moves, both its position and its orientation vary with time. The translation and rotation of the body are measured with respect to a reference coordinates system. In a rigid body, all particles of the body move with the same translational velocity. However, when rotating, all particles of the body (except those lying on the axis of rotation) change their position.

Thus, the linear velocity of any two particles of the body may not be the same. Angular velocities of all particles are the same.

When an object has translational and/or rotational motion, the radar scattering from the object is subject to modulation in its amplitude and phase. Theoretical analysis indicates that the motion of an object can modulate the phase function of the scattered electromagnetic (EM) waves. If the object oscillates periodically, the modulation generates sideband frequencies about the frequency of the incident wave.

To incorporate any object's motion in EM simulation, first, the trajectory and orientation of the object must be determined by using motion differential equations and a rotation matrix of the object. Then, using the quasi-static method, the motion of the object is considered a sequence of snapshots taken at each time instant. Finally, using a suitable RCS prediction method, the scattering EM field is estimated.

The simplest model of the EM scattering mechanism is the point scatterer model. With this model, an object is defined in terms of a 3-D reflectivity density function characterized by point scatterers. It is rather straightforward to incorporate an object's motion in the point scatterer model simulation.

Another simple method of radar cross-section (RCS) modeling is to decompose an object into canonical geometric components, such as a sphere, an ellipsoid, or a cylinder. The RCS of each canonical component can be expressed by a mathematical formula.

A more accurate RCS modeling method is the physical optics (PO) model. It is a simple and convenient RCS prediction method for any 3-D object.

In this chapter, typical examples of an oscillating pendulum, rotating rotor blades, a spinning symmetric top, and wind turbines are introduced. Pendulum oscillation is a commonly used example to understand the basic principle of the nonlinear motion dynamic. Rotating helicopter rotor blades are one of the most popular subjects in radar target signature analysis. The spinning symmetric top has more complex nonlinear motion and has shown more interesting signatures than other rigid body motion. Wind turbines have now become challenges to current radar systems, and large numbers of wind farms and the large RCS of turbine blades have significant impacts on radar performance. In this chapter, details of the modeling of a nonlinear motion dynamic, the modeling of the RCS of a rigid body, the mathematical model of radar scattering from a rotating rigid body, and the micro-Doppler signature of a typical rigid body will be described, and the simulation of nonlinear motion and radar backscattering using MATLAB will be provided.

3.1 Pendulum Oscillation

Pendulum oscillation is a commonly used example to understand the basic principle of nonlinear motion dynamic. A simple pendulum is modeled by a weighted small bob, attached to one end of a weightless string, with the other end of the string fixed to a pivot point as shown in Figure 3.1. Under the influence of gravity $g = 9.80665$ m/s^2, the small bob swings back and forth periodically about a fixed horizontal axis along the y-axis and at ($x = 0$, $z = 0$). In a stable equilibrium condition, the center of mass of the pendulum is right below the axis of rotation at ($x = 0$, $y = 0$, $z = L$), where L is the length of the string.

3.1.1 Modeling Nonlinear Motion Dynamic of a Pendulum

Newton's second law states that the total force acting on the pendulum is equal to the product of the mass of the pendulum and its acceleration. If the pendulum is initially deviated from its stable position by a swinging angle θ, there are two forces acting on the mass of the pendulum: the downward gravitational force, mg, where m is the mass of the pendulum and g is the gravitational acceleration, and the tension, T, in the string. However, the

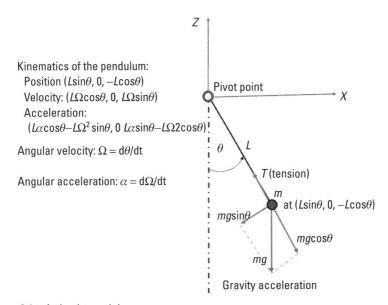

Figure 3.1 A simple pendulum.

tension has no contribution to the torque because its line of action passes through the pivot point. From simple trigonometry, the line of action of the gravitational force passes a distance $L\sin\theta$ from the pivot point. Hence, the magnitude of the gravitational torque is $mgL\sin\theta$. The gravitational torque is a restoring torque (i.e., if the mass of the pendulum is displaced slightly from its equilibrium state at $\theta = 0$, then the gravitational force acts to pull the pendulum back toward that state).

According to the Newton's second law of motion, the net torque is $I(d^2\theta/d^2t)$, where I is the moment of inertia, and $(d^2\theta/d^2t)$ is the angular acceleration. If the torque acting on the system is τ, then $\tau = I(d^2\theta/d^2t)$ is the angular equation of motion. Given the length of the string L and the mass of the pendulum m, the moment of inertia about the pivot is

$$I = mL^2 \tag{3.1}$$

The vector torque $\boldsymbol{\tau}$ is the cross-product of the position vector $\boldsymbol{L}$ and the gravitational force vector $m\boldsymbol{g}$ (i.e., $\boldsymbol{\tau} = \boldsymbol{L} \times m\boldsymbol{g}$). The magnitude of the torque is

$$\tau = Lmg\sin\theta \tag{3.2}$$

and the net torque on the pendulum is

$$-Lmg\sin\theta = I\frac{d^2\theta}{dt^2} = mL^2\frac{d^2\theta}{dt^2} \tag{3.3}$$

and, finally, the equation of pendulum becomes

$$mL\frac{d^2\theta}{dt^2} = -mg\sin\theta \tag{3.4}$$

This equation defines the relationship between the swinging angle θ and its second time derivatives $d^2\theta/dt^2$.

By denoting the angular velocity $\Omega = d\theta/dt$, the equation of pendulum can be rewritten as a set of two first-order ordinary differential equations (ODEs):

$$\begin{cases} \dfrac{d\theta}{dt} = \Omega \\ \dfrac{d\Omega}{dt} = -\dfrac{g}{L}\sin\theta \end{cases} \tag{3.5}$$

Figure 3.2 shows the oscillating angle and angular velocity of a simple pendulum with the mass $m = 20\text{g}$ and the length $L = 1.5\text{m}$.

For small angle θ, $\sin\theta$ can be substituted by θ and the pendulum becomes a linear oscillator. Thus, the differential equation of the pendulum motion becomes

$$\frac{d^2\theta}{dt^2} + \omega_0^2\theta \cong 0 \tag{3.6}$$

where $\omega_0 = (g/L)^{1/2}$ is the angular frequency of the oscillating pendulum. Equation (3.6) is a harmonic equation and the solution of the swinging angle is

$$\theta(t) = \theta_0 \sin\omega_0 t \tag{3.7}$$

and its angular velocity is

$$\Omega(t) = \frac{d\theta(t)}{dt} = \theta_0\omega_0 \cos\omega_0 t \tag{3.8}$$

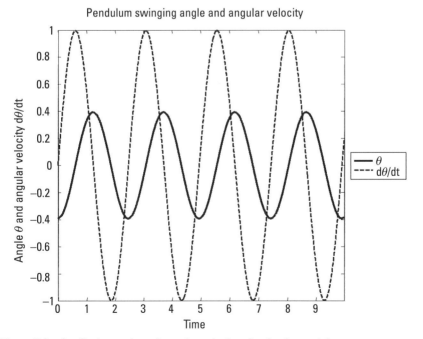

Figure 3.2 Oscillating angle and angular velocity of a simple pendulum.

where θ_0 is the initial swinging angle of the pendulum, called the initial amplitude.

For a given initial amplitude θ_0, the period of the oscillating pendulum is determined by

$$T_0 = 2\pi \frac{1}{\omega_0} = 2\pi \sqrt{\frac{L}{g}} \left(1 + \frac{1}{4} \sin^2 \frac{\theta_0}{2} + \frac{9}{64} \sin^4 \frac{\theta_0}{2} + \dots \right) \qquad (3.9)$$

For a small initial amplitude θ_0, the oscillating period is

$$T_0 = 2\pi \sqrt{\frac{L}{g}} \qquad (3.10)$$

or the frequency of the oscillation is $f_0 = 1/T_0$.

The simple pendulum assumes that the string is weightless and the bob is small such that its angular momentum is negligible. However, a physical pendulum may have a large size and mass. Thus, it may have a significant moment of inertia I.

From (3.3), the equation of a physical pendulum can be written as

$$I \frac{d^2\theta}{dt^2} = -mgL_{\text{effect}} \sin\theta \qquad (3.11)$$

where L_{effect} is the effective length of a physical pendulum, and the right side of the equation is the net torque of the gravity. The physical pendulum equation can be simply expressed as

$$\frac{d^2\theta}{dt^2} + \omega_0^2 \sin\theta = 0 \qquad (3.12)$$

where $\omega_0^2 = mgL_{\text{effect}}/I$ is the angular frequency of the physical pendulum. Thus, the period of its swinging becomes

$$T_0 = \frac{2\pi}{\omega_0} = 2\pi \sqrt{\frac{I}{mgL_{\text{effect}}}} \qquad (3.13)$$

The physical pendulum equation with an effective length L_{effect} is the same as the simple pendulum with its length $L = L_{\text{effect}}$. The physical pendulum (3.12) is described by the same mathematic formula and is equivalent to the simple pendulum.

If linear friction exists in the oscillating pendulum, an additional term, $-2\gamma(d\theta/dt)$, proportional to the angular velocity, must be added to the right side of (3.4). Then the equation of pendulum becomes

$$\frac{d^2\theta}{dt^2} + 2\gamma\frac{d\theta}{dt} + \omega_0^2\sin\theta = 0 \tag{3.14}$$

where $\omega_0 = (g/L)^{1/2}$ is the angular frequency of free oscillations, and γ is the damping constant. Thus, the equation of pendulum with linear friction can be rewritten as a set of two first-order ODEs

$$\begin{cases} \dfrac{d\theta}{dt} = \Omega \\ \dfrac{d\Omega}{dt} + 2\gamma\Omega = -\dfrac{g}{L}\sin\theta \end{cases} \tag{3.15}$$

For a small angle θ, $\sin\theta \approx \theta$ and the pendulum equation is approximately

$$\frac{d^2\theta}{dt^2} + 2\gamma\frac{d\theta}{dt} + \omega_0^2\theta = 0 \tag{3.16}$$

If the friction is weak such that $\gamma < \omega_0$, the solution of (3.16) is

$$\theta(t) = \theta_0 e^{-\gamma t}\cos\left(\omega t + \varphi_0\right) \tag{3.17}$$

where θ_0 is the initial amplitude, φ_0 is the initial phase depending on the initial excitation, and the exponential term $\theta_0\exp(-\gamma t)$ is a decreasing factor. The angular frequency of the oscillation ω is given by $\omega = \sqrt{\omega_o^2 - \gamma^2}$ $= \omega_0\sqrt{1 - (\gamma/\omega_0)^2}$. When $\gamma < \omega_0$, the angular oscillation frequency and period become

$$\omega \approx \omega_0 - \gamma^2/(2\omega_0)$$
$$T \approx T_0\left[1 + \gamma^2/(2\omega_0^2)\right] \tag{3.18}$$

which are close to the free oscillation frequency ω_0 and period T_0.

Figure 3.3(a) shows the oscillating angle and angular velocity of the damping pendulum with damping constant $\gamma = 0.07$ and $\omega_0 = (g/L)^{1/2} = 2.56$.

If there is a damping effect as well as a driving force in the pendulum oscillation, the equation of the pendulum must be modified to

$$\frac{d^2\theta}{dt^2} + 2\gamma\frac{d\theta}{dt} + \frac{g}{L}\sin\theta = \frac{A_{Dr}}{mL}\cos\left(2\pi f_{Dr}t\right) \tag{3.19}$$

where γ is the damping constant, A_{Dr} is the amplitude of the driving, and f_{Dr} is the driving frequency.

If we let the angular velocity $\Omega = d\theta/dt$, the equation of pendulum with friction and driving force can be rewritten as a set of two first-order ODEs

$$\begin{cases} \dfrac{d\theta}{dt} = \Omega, \\ \dfrac{d\Omega}{dt} + 2\gamma\Omega = -\dfrac{g}{L}\sin\theta + \dfrac{A_{Dr}}{mL}\cos\left(2\pi f_{Dr}t\right) \end{cases} \tag{3.20}$$

Figure 3.3(b) shows the oscillating angle and angular velocity of the damping driving pendulum with damping constant $\gamma = 0.07$, driving amplitude $A_{Dr} = 15$, and normalized driving frequency $f_{Dr} = 0.2$.

3.1.2 Modeling RCS of a Pendulum

The RCS measures the strength of an object's reflectivity and is a function of the object's orientation and the radar transmitted frequency. The small bob of a pendulum has a simple geometric shape, such as a sphere, an ellipsoid, or a cylinder, and the RCS of a simple geometric shape can be expressed by a mathematical formula.

High-frequency RCS prediction methods and exact RCS prediction formulas can be found in the book [1] by Knott, Shaeffer, and Tuley. The computer simulation of radar backscattering used in this book is not an exact RCS prediction. Instead, the simulation is based on approximate and simplified complex scattering solutions [2]. With the simplest component method, an object consists of a limited number of the simplest components, such as spheres, ellipsoids, and cylinders. The formulas of the simplest components are available, but are not exact solutions.

The RCS of a perfectly conducting sphere has three regions. In the optical region, which corresponds to a sphere that is large compared with the wavelength, the RCS is a constant and can be simply expressed by $RCS_{sphere} = \pi r^2$, where r is the radius of the sphere and is much greater than the wavelength λ. In the Rayleigh region for a small sphere, the RCS is $RCS_{sphere} = 9\pi r^2 (kr)^4$,

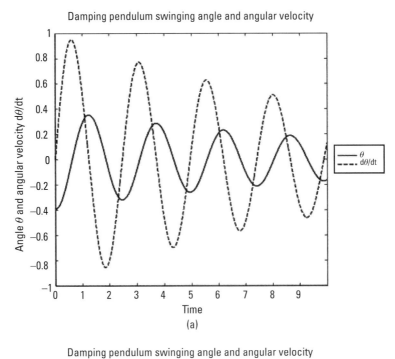

Damping pendulum swinging angle and angular velocity
(a)

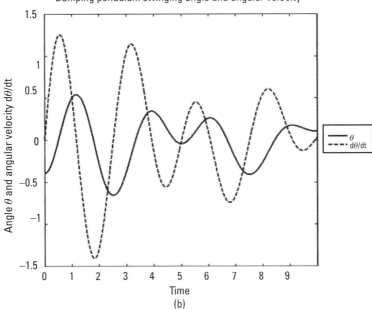

Damping pendulum swinging angle and angular velocity
(b)

Figure 3.3 Oscillating angle and angular velocity of: (a) a damping pendulum, and (b) a damping driving pendulum.

where $k = 2\pi/\lambda$. The region between the Rayleigh and optical regions is a resonance region called the Mie region [1, 3].

An approximation for the RCS of an ellipsoid backscattering is given by [3]

$$RCS_{\text{ellip}} = \frac{\pi a^2 b^2 c^2}{\left(a^2 \sin^2\theta \cos^2\varphi + b^2 \sin^2\theta \sin^2\varphi + c^2 \cos^2\theta\right)^2} \quad (3.21)$$

where a, b, and c represent the length of the three semi-axes of the ellipsoid in the x, y, and z directions, respectively. The incident aspect angle θ and the azimuth angle φ represent the orientation of the ellipsoid relative to the radar, and defined by

$$\theta = \arctan\left(\frac{\sqrt{x^2 + y^2}}{z}\right) \quad (3.22)$$

and

$$\varphi = \arctan\left(\frac{y}{x}\right) \quad (3.23)$$

where the incident angle counts from the z-axis and the azimuth angle counts from the x-axis. If the ellipsoid is symmetric (i.e., $a = b$), the RCS will be independent of the azimuth angle φ.

The nonnormal incidence backscattered RCS for a symmetric cylinder due to a linear polarized incident wave is approximated by [3]

$$RCS_{\text{cylinder}} = \frac{\lambda r \sin\theta}{8\pi \cos^2\theta} \quad (3.24)$$

where r is the radius, θ is the incident aspect angle, and the RCS is independent of the azimuth angle φ.

These RCS formulas can be used to simulate radar backscattering from an oscillating pendulum.

3.1.3 Radar Backscattering from an Oscillating Pendulum

To calculate radar backscattering from an oscillating pendulum, ordinary differential equations are used for solving the swinging angle and the angular velocity. Therefore, at each time instant during a radar observation time interval, the location of the pendulum can be determined. Based on the location and orientation of the pendulum, the RCS of the pendulum and the radar received signal can be calculated.

If the radar transmits a sequence of narrow rectangular pulses with a transmitted frequency f_c, a pulse width Δ, and a pulse repetition interval ΔT, the radar received baseband signal is

$$s_B(t) = \sum_{k=1}^{n_p} \sqrt{\sigma_P(t)}\, rect\left\{t - k\Delta T - \frac{2R_P(t)}{c}\right\} \exp\left\{-j2\pi f_c \frac{2R_P(t)}{c}\right\} \qquad (3.25)$$

where $\sigma_P(t)$ is the RCS of the small bob at time t, n_p is the total number of pulses received, $R_P(t)$ is the distance from the radar to the small bob at time t, and the rectangular function *rect* is defined by

$$rect(t) = \begin{cases} 1 & 0 \leq t \leq \Delta \\ 0 & \text{otherwise} \end{cases} \qquad (3.26)$$

Given the location of the radar at $(x = 10\text{m}, y = 0\text{m}, z = 0\text{m})$, the pivot point of the pendulum is assumed at $(x = 0\text{m}, y = 0\text{m}, z = 2\text{m})$. The string length $L = 1.5\text{m}$ and the mass of the small bob is 20g. In cases of damping and driving, let the damping constant be $\gamma = 0.07$, and the driving amplitude be $A_{Dr} = 15$, and the normalized driving frequency be $f_{Dr} = 0.2$. The geometric configuration of the radar and the pendulum is illustrated in Figure 3.4.

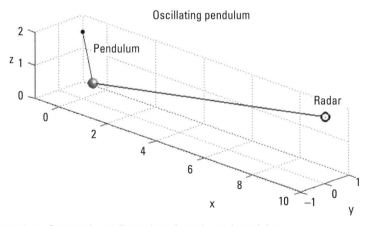

Figure 3.4 Geometric configuration of a radar and pendulum.

Equations (3.5), (3.15), and (3.20) are used to calculate the oscillating angle and angular velocity of the simple, the damping, and the damping and driving pendulum, respectively. In the rotation matrix of the pendulum, only the pitch angle varies and the roll and yaw angles are always zero. The RCS of the small bob is simulated by the point-scatterer model because the small bob can be seen as a point scatterer. After arranging the n_p range profiles, the two-dimensional (2-D) pulse-range profiles can be obtained. Figure 3.5(a) shows the 2-D range profiles of the simple oscillating pendulum, where the radar wavelength is 0.03m at the X-band. The oscillating small bob can be seen around a distance of 10m from the radar.

3.1.4 Micro-Doppler Signatures Generated by an Oscillating Pendulum

The micro-Doppler signature of the oscillating simple pendulum, shown in Figure 3.5(b), is obtained from the summation of those range profiles that are within a range gate around 10m, where the small bob is located within the radar observation interval. The joint time-frequency transform used to generate the signature is a simple short-time Fourier transfer (STFT). Other higher-resolution time-frequency transforms, such as the smoothed pseudo-Wigner-Ville distribution, may also be used.

Compared with the micro-Doppler signature of a simple pendulum, Figure 3.6 shows the micro-Doppler signatures of a damping pendulum and a damping and driving pendulum with $L = 1.5$m, $m = 20$g, $\gamma = 0.07$, $A_{Dr} = 15$, and $f_{Dr} = 0.2$.

From the micro-Doppler signature in Figure 3.6(a), an oscillating frequency of 0.4 Hz can be measured. The damping constant γ is measured from the change of the amplitude of the Doppler modulation during the observation time duration of 10 seconds. Due to the measured change of the amplitude of the Doppler modulation being 101 Hz/202 Hz during a 10-second time interval, the damping constant is estimated as

$$\gamma = -\log_e(101/202)/10 = 0.069$$

which is consistent with the damping constant of 0.07 used in the simulation. The MATLAB code for calculating radar backscattering from an oscillating pendulum is provided in this book.

Range profiles of an oscillating pendulum

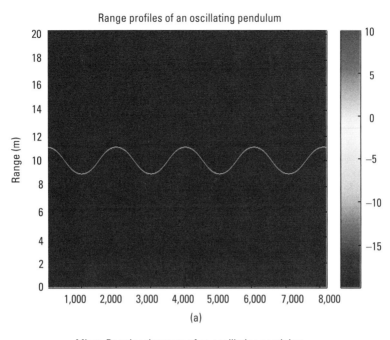

(a)

Micro-Doppler signature of an oscillating pendulum

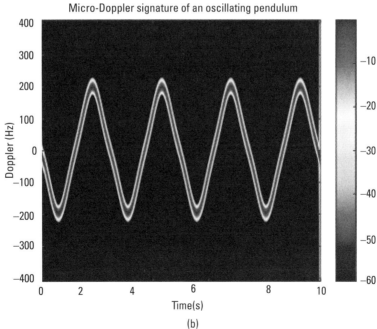

(b)

Figure 3.5 (a) Range profiles, and (b) micro-Doppler signature of a simple free oscillating pendulum.

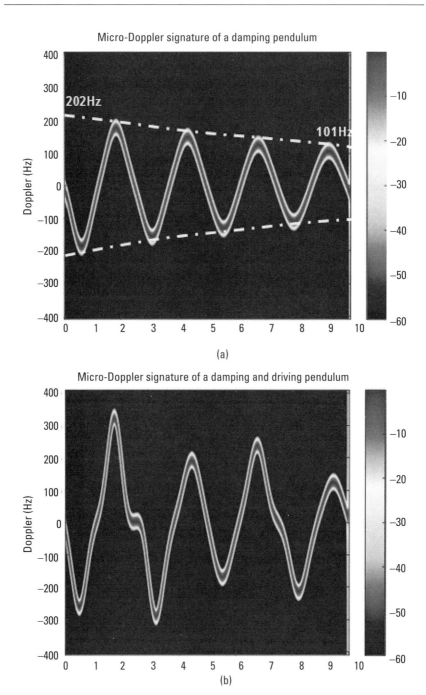

Figure 3.6   The micro-Doppler signatures of (a) a damping oscillating pendulum, and (b) a damping and driving pendulum.

3.2   Helicopter Rotor Blades

An airfoil of a helicopter rotor blade and its cross-section profile is shown in Figure 3.7. Different types of airfoils have various shapes and dimensions. A rotating aerofoil always has bending, flexing, and twisting. However, in the simulation study of a helicopter's rotor blades, no bending, flexing, and twisting are considered.

Blades of helicopters are usually metallic or a composite material that produces strong radar reflectivity. EM scattering from an airfoil mainly includes specular reflections from its surfaces and leading edge, diffraction from its trailing edge, creeping waves around the leading edge, and traveling waves from the trailing edge. Radar returns from a helicopter have its unique spectral signature [4–6]. Figure 3.8 illustrates a general spectral signature of helicopters with rotating rotor blades. The spectral signature has spectral components from the fuselage, from the rotor hub, from the main rotor's receding blades, and from approaching blades. The strongest spectral amplitude comes from the fuselage. The spectral amplitude of the receding blade is different from that of the approaching blade because of the difference between the leading edge and the trailing edge. Among these spectral features, the receding blades and the approaching blades are especially interesting.

3.2.1   Mathematic Model of Rotating Rotor Blades

The geometry of the radar and rotating rotor blades is shown in Figure 3.9. The radar is located at the origin of the space-fixed coordinates (X, Y, Z) and the rotor blades are centered at the origin of the body-fixed coordinates on the plane $(x, y, z = 0)$ rotating about the z-axis with an angular rotation rate Ω. The reference coordinates (X', Y', Z') is parallel to and translated from the space-fixed coordinates located at the same origin as the body-fixed coordinates.

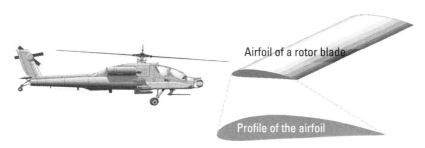

Figure 3.7   An example of the airfoil of a helicopter's rotor blade.

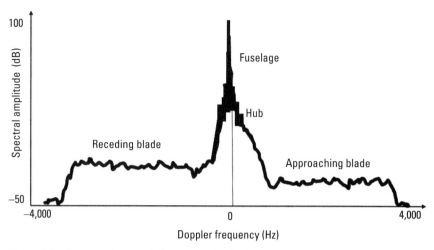

Figure 3.8 A general spectral signature of radar back-scattering from a helicopter.

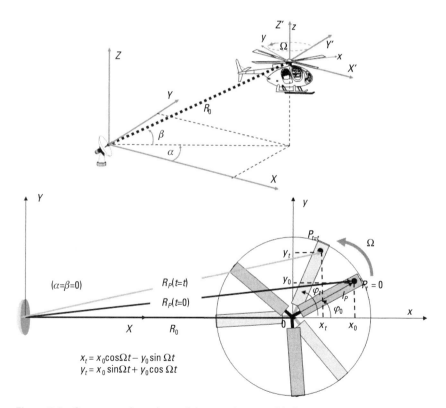

Figure 3.9 Geometry of a radar and the rotating rotor blades.

The distance from the radar to the origin of the reference coordinates is R_0. The radar observed azimuth and elevation angles of the origin of the reference coordinates are α and β, respectively.

From the EM scattering point of view, each blade of the rotor consists of scatterer centers. Each scatterer center is considered a point with a certain reflectivity. For simplicity, the same reflectivity is assigned to all of the scatterer centers. Let $\alpha = \beta = 0$; if a point scatterer P at $(x_0, y_0, z_0 = 0)$ rotates about the z-axis in the body-fixed coordinates with a constant angular rotation rate Ω, the distance from the origin of the body-fixed coordinates to the point scatterer P is $l_P = (x_0^2 + y_0^2)^{1/2}$. If the initial rotation angle of the point P at $t = 0$ is φ_0, then at time t the rotation angle becomes $\varphi_t = \varphi_0 + \Omega t$ and the point P rotates to $(x_t, y_t, z_t = 0)$ as shown in Figure 3.9. Thus, the range from the radar to the point scatterer P becomes

$$R_P(t) = \left[R_0^2 + l_P^2 + 2l_P R_0 \cos(\varphi_0 + \Omega t) \right]^{1/2}$$
$$\cong R_0 + l_P \cos\varphi_0 \cos\Omega t - l_P \sin\varphi_0 \sin\Omega t \qquad (3.27)$$

where assuming $(l_P/R_0)^2 \to 0$ in the far field. Then the radar received signal returned from the point scatterer P is

$$s_R(t) = \exp\left\{ -j\left[2\pi f t + \frac{4\pi}{\lambda} R_P(t) \right] \right\} = \exp\left\{ -j\left[2\pi f t + \Phi_P(t) \right] \right\} \qquad (3.28)$$

where $\Phi_P(t) = 4\pi R_P(t)/\lambda$ is the phase function of the point scatterer and assuming the RCS of the point scatterer $\sigma_P = 1$.

If the elevation angle β and the height z_0 of the rotor blades are not zero, the phase function can be modified as

$$\Phi_P(t) = \frac{4\pi}{\lambda}\left[R_0 + \cos\beta\left(l_P \cos\varphi_0 \cos\Omega t - l_P \sin\varphi_0 \sin\Omega t \right) + z_0 \sin\beta \right] \quad (3.29)$$

and, thus, the returned signal from the point scatterer P becomes

$$s_R(t) = \exp\left\{ j\frac{4\pi}{\lambda}\left[R_0 + z_0 \sin\beta \right] \right\} \exp\left\{ -j2\pi f t - \frac{4\pi}{\lambda} l_P \cos\beta \cos\left(\Omega t + \varphi_0 \right) \right\}$$

$$(3.30)$$

In a similar way to (2.76), (2.77), and (2.78) in Chapter 2, denoting $B = (4\pi/\lambda)l_P \cos\beta$, (3.30) can be expressed in terms of the Bessel function $J_k(B)$ of first kind of order k. Thus, the spectrum of the point scatterer P consists of pairs of spectral lines around the center frequency f and with a spacing $\Omega/(2\pi)$ between adjacent lines [7].

The baseband signal returned from the point scatterer P is a monocomponent signal:

$$s_B(t) = \exp\left\{-j\frac{4\pi}{\lambda}\left[R_0 + z_0\sin\beta\right]\right\}\exp\left\{-j\frac{4\pi}{\lambda}l_P\cos\beta\cos(\Omega t + \varphi_0)\right\} \quad (3.31)$$

By integrating (3.31) over the length of the blade L, the total baseband signal returned from one blade becomes [8, 9]

$$
\begin{aligned}
s_L(t) &= \exp\left\{-j\frac{4\pi}{\lambda}\left[R_0 + z_0\sin\beta\right]\right\}\int_0^L \exp\left\{-j\frac{4\pi}{\lambda}l_P\cos\beta\cos(\Omega t + \varphi_0)\right\}dl_P \\
&= L\exp\left\{-j\frac{4\pi}{\lambda}\left[R_0 + z_0\sin\beta\right]\right\}\exp\left\{-j\frac{4\pi}{\lambda}\frac{L}{2}\cos\beta\cos(\Omega t + \varphi_0)\right\} \\
&\quad \text{sinc}\left\{\frac{4\pi}{\lambda}\frac{L}{2}\cos\beta\cos(\Omega t + \varphi_0)\right\}
\end{aligned}
$$

$$(3.32)$$

where sinc($\cdot$) is the sinc function: sinc(x) = 1 when x = 0; sinc(x) = sin(x)/x when $x \neq 0$.

For a rotor with N blades, the N blades have N different initial rotation angles:

$$\theta_k = \theta_0 + k2\pi/N, (k = 0,1,2,\ldots N-1)$$

and the total received baseband signal returned from the rotor is

$$s_\Sigma(t) = \sum_{k=0}^{N-1} s_{Lk}(t) = L\exp\left\{-j\frac{4\pi}{\lambda}\left[R_0 + z_0\sin\beta\right]\right\}$$

$$\sum_{k=0}^{N-1}\text{sinc}\left\{\frac{4\pi}{\lambda}\frac{L}{2}\cos\beta\cos\left(\Omega t + \varphi_0 + k\frac{2\pi}{N}\right)\right\}\exp\left\{-j\Phi_k(t)\right\} \quad (3.33)$$

where the phase function

$$\Phi_k(t) = \frac{4\pi}{\lambda}\frac{L}{2}\cos\beta\cos(\Omega t + \varphi_0 + k2\pi/N) \quad (k = 0,1,2,\ldots N-1) \quad (3.34)$$

The time-domain signature of the rotor blades is given by the magnitude of (3.33):

$$|s_\Sigma(t)| = \left| L \exp\left\{ -j\frac{4\pi}{\lambda}\left[R_0 + z_0\sin\beta \right] \right\} \right.$$

$$\left. \sum_{k=0}^{N-1} \text{sinc}\left\{ \frac{4\pi}{\lambda}\frac{L}{2}\cos\beta\cos\left(\Omega t + \varphi_0 + k\frac{2\pi}{N} \right) \right\} \exp\left\{ -j\Phi_k(t) \right\} \right|$$

(3.35)

In (3.33), the total received baseband signal returned from the rotor is a multicomponent signal. As discussed in Section 1.8, the instantaneous frequency analysis is not suitable for multicomponent signals. To deal with the multicomponent signal, the complete time-frequency distribution can be obtained by first computing the time-frequency distribution for each mono-component signal returned from each point scatterer and then combining these individual time-frequency distributions together.

Thus, based on the point-scatterer model and without calling the sinc function, first, each blade can be simplified as a combination of multiple point scatterers. Then the baseband signal returned from one blade is modeled by the sum of the returns from all point scatterers:

$$s_\Sigma(t) = \sum_{k=0}^{N-1} \exp\left\{ -j\frac{4\pi}{\lambda}\left[R_0 + z_0\sin\beta \right] \right\} \sum_{p=1}^{N_p} \exp\left\{ -j\Phi_{k,p}(t) \right\} \quad (3.36)$$

where N_p is the total number of scatterers in each blade and the phase function is

$$\Phi_{k,p}(t) = \frac{4\pi}{\lambda}l(p)\cos\beta\cos\left(\Omega t + \varphi_0 + \frac{k2\pi}{N} \right) \quad \left(k = 0,1,\dots N-1;\ p=1,\dots N_p \right)$$

From (3.33) to (3.36), the frequency spectrum of rotating rotor blades can be estimated. In (3.33), the integration over the length of the blade in (3.36) is represented by the sinc function.

Assume the radar is C-band with a wavelength of $\lambda = 0.06$m and the target is a helicopter. The main rotor of the helicopter has two blades rotating with a constant rotation rate $\Omega = 4$ revolutions per second (r/s) (or $4 \times 2\pi$ rad/sec). The length of the blade from the rotor center to the blade tip is $L = 6.5$m. The main rotor of the helicopter is at distance of 700m from the radar to the center of the rotor with an elevation angle $\beta = 45°$. Thus, the tangential velocity of the rotating blade tip is determined by the rotation rate Ω and the length of the blade tip L: $V_{tip} = 2\pi L\Omega = 163.4$ m/s. The maximum Doppler shift becomes $\{f_D\}_{max} = (2V_{tip}/\lambda) \cos\beta = 3.85$ kHz and the Nyquist rate is two times of the maximum Doppler shift $2 \times \{f_D\}_{max} = 7.7$ kHz. If the digital sampling rate is 10 kHz, there is no frequency aliasing. The time-domain signature of the rotor blades is calculated as shown in Figure 3.10(a) and the frequency spectrum of the same signal is shown in Figure 3.10(b).

The radar returned signal from the rotor blades has short flashes when a blade has a specular reflection at the approaching or advancing points and the receding points [9]. The time interval between two successive flashes is related to the rotation rate of the rotor. The width of the flash is determined by the blade length L, the wavelength λ, the elevation angle β, and the rotation rate Ω as described by the sinc function in (3.33). For a longer blade length and at a shorter wavelength, the width of the flash is shorter. Because the number

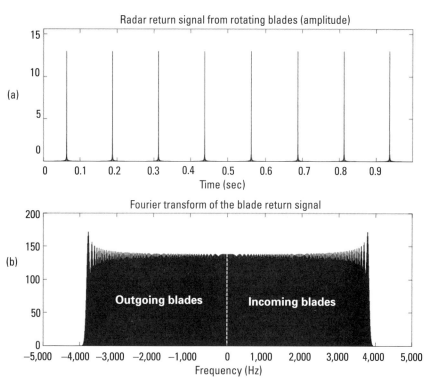

Figure 3.10 (a) Time-domain signature of a two-blade rotor, and (b) the frequency spectrum of a two-blade rotating rotor.

of blades is $N = 2$ and the rotation rate is $\Omega = 4$ r/s, there are 8 flashes in 1.0 second for each of the blades, and the interval between flashes is $T_{\text{flash}} = 1/8 = 0.125$ (second) as shown in Figure 3.10(a).

Because rotating rotor blades impart periodic modulations on the radar returned signal, the rotation-induced Doppler shifts occupy unique locations in the frequency domain relative to the Doppler shift of the fuselage. Figure 3.10(b) shows the spectral components (outgoing blades and incoming blades) of the two-blade rotor without fuselage and rotor hub. The Doppler shift of the fuselage and the hub are around zero Doppler if the helicopter has no translational motion. The total width of the Doppler spectrum is 2 times of the maximum Doppler frequency, $2\{f_D\}_{\max} = 7.7$ kHz. The interval of frequency sampling is the number of blades times of the rotation rate, $N \times \Omega = 8$ Hz.

The rotation feature of rotor blades is considered an important feature for identifying helicopters of interest [9, 10]. The Doppler modulation induced by rotating rotor blades is regarded as a unique signature of helicopters. Representing the Doppler modulation in the joint time-frequency domain, the micro-Doppler signature of the rotor blades can be seen. Figure 3.11 is the

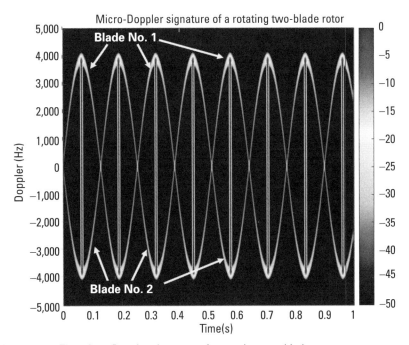

Figure 3.11   The micro-Doppler signature of a rotating two-blade rotor.

micro-Doppler signature of a rotating two-blade rotor based on (3.33), where 8 flashes from the blade no. 1 and other 8 flashes from the blade no. 2 can be seen. There is no strong zero-Doppler point at these flashes due to 180° phase cancel out in the two-blade case.

For comparison, Figure 3.12 shows the micro-Doppler signature of a rotating three-blade rotor, where each blade has 8 flashes and the total number of fleshes is 24. A strong zero-Doppler zone can be seen for the three-blade case. In cases when the rotor has even number of blades, the zero-Doppler zone disappears from micro-Doppler signatures. This is because the pair of blades that have a 180° difference between their rotating angles cancels out zero-Doppler components.

Based on the above simulation studies, the micro-Doppler signature of rotating blades shows the following features:

1. The width of the Doppler spectrum, $2\{f_D\}_{max}$, is a function of rotation rate Ω, blade length L, elevation angle β, and the wavelength λ (i.e., $\{f_D\}_{max} = (4\pi L\Omega/\lambda)\cos\beta$). Thus, longer blade length, faster rotation rate, and shorter wavelength can widen the Doppler spectrum.

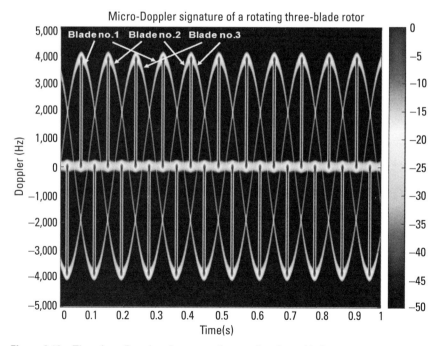

Figure 3.12   The micro-Doppler signature of a rotating three-blade rotor.

2. The micro-Doppler signature of each blade is a sinusoidal function at a frequency of the rotation rate Ω (i.e., $\{f_D\}_{max} \cdot \sin(\Omega t + k2\pi/N)$, ($k = 1, 2, \dots N$)), with an initial phase different from other blades.
3. Due to specular scattering when a blade is normal to the radar line of sight (LOS), the flash occurs and shows a series of strong scattered fields with maximum Doppler shifts. These flashed lines are located at the time instant $t = kT_{flash}/2$, where k is an odd integer ($k = 1, 3, 5, 7, 9, \dots$) and $T_{flash} = \pi/\Omega$.
4. There is a strong zero-Doppler zone in the micro-Doppler signature, which is generated by the scattering from the blade sections close to the center of rotation.

3.2.2 RCS Model of Rotating Rotor Blades

To calculate EM scattering from rotating rotor blades, for simplicity, the blade in Figure 3.7 is simplified as a rigid, homogeneous, linear rectangular flat plate rotating about a fixed axis with a constant rotation rate and without considering the leading edge and the trailing edge. No flapping, lagging, and feathering are considered in the rectangular flat. The geometry of the rotor blade and the radar is illustrated in Figure 3.13. For a perfectly-conducting rectangular flat plate, the mathematical formula of the RCS can be found in [3, 11]. An approximation for the RCS of a rectangular flat plate is given by

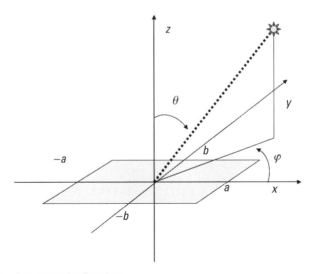

Figure 3.13 A rectangular flat plate.

[3]. In the RCS formula, there are two terms: the peak RCS σ_{Peak} and the aspect factor σ_{Aspect}:

$$\sigma = \sigma_{\text{Peak}}\sigma_{\text{Aspect}} = \frac{4\pi a^2 b^2}{\lambda^2}\left(\cos\theta\frac{\sin x_k}{x_k}\frac{\sin y_k}{y_k}\right)^2 \tag{3.37}$$

where $\sigma_{\text{Peak}} = 4\pi a^2 b^2/\lambda^2$, $\sigma_{\text{Aspect}} = (\cos\theta(\sin x_k/x_k)(\sin y_k/y_k))^2$, $x_k = ka\sin\theta\sin\varphi$, $y_k = kb\sin\theta\cos\varphi$, and $k = 2\pi/\lambda$. Equation (3.37) is independent of the polarization and is accurate only for small aspect angle $\theta \leq 20°$.

3.2.3 POFACET Prediction Model

Physical optics (PO) is a convenient method for predicting the RCS of any 3-D object. It is a high-frequency region (or optical region) prediction and provides the best results for objects with a dimension much larger than the wavelength. The PO method applies to the illuminated surfaces, but does not apply edge diffractions, multiple reflections, or surface waves.

Any complex larger surface can be divided into many small surfaces, called facets. The facet used in the POFACET model is a triangular flat plate. The scattered field from each facet can be calculated as if it were isolated without considering the effect of other facets. Thus, for a facet illuminated by the incident field, its surface current and scattered field can be calculated. For a shadowed surface, its surface current is set to zero.

Based on the incident and scattered fields, the RCS of a surface is determined by $\sigma = \lim_{R\to\infty} 4\pi R^2(|E_s|^2/|E_i|^2)$, where R is the range from the radar to the surface, and $|E_s|$ and $|E_i|$ are the amplitudes of the scattered and the incident electric fields, respectively.

As shown in Figure 3.14, an incident wave is described by the spherical coordinate angles θ and φ. The polarization of an incident wave can be decomposed into two orthogonal components in terms of the angles θ and φ to represent the incident field in the spherical system. Thus, the incident field is represented by $E_i = E_\theta n_\theta + E_\varphi n_\varphi$, where n_θ and n_φ are the unit vectors in the spherical system.

For calculating the EM scattering from a facet defined by three vertices given by points P_1, P_2, and P_3 with an arbitrary orientation, the body-fixed coordinate system is selected such that the triangular facet lies on the (x, y)-plane and the direction of its unit normal vector n is identical to the z-axis, as shown in Figure 3.15.

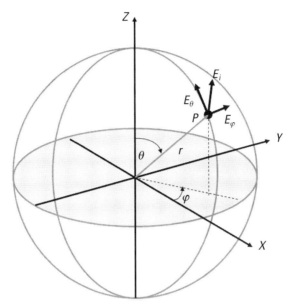

Figure 3.14 The spherical coordinate system used in POFACET computation.

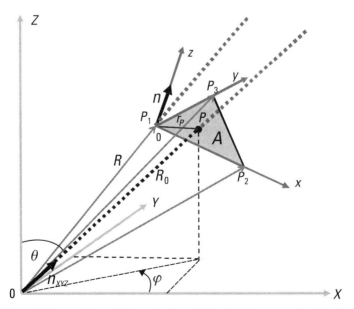

Figure 3.15 Arbitrary oriented triangular facet defined in the space-fixed system (X, Y, Z) and in the body-fixed local system (x, y, z).

Because the radar is in the far zone and the size of the object is much smaller than the distance R_0 from the radar to the object, the range vector $\boldsymbol{R}$ from the radar to the origin of the body-fixed coordinate (x, y, z) can be considered to be parallel to the vector $\boldsymbol{R_0}$. In the space-fixed coordinates, the unit vector of the radar LOS is $\boldsymbol{n}_{XYZ} = [u, v, w]$, where $u = \sin\theta\cos\varphi$, $v = \sin\theta\sin\varphi$, and $w = \cos\theta$. In the body-fixed coordinates, any point P located at (x_P, y_P, z_P) on the facet is represented by its position vector $\boldsymbol{r}_P = [x_P, y_P, z_P]$. Thus, the scattered field from the facet is given by [12]

$$E_S(R,\theta,\varphi) = \frac{-jk\mathfrak{R}_{\text{imp}}}{4\pi R}\exp(-jkR)\iint_A J_s \exp\left[jk\left(\boldsymbol{r}_P \cdot \boldsymbol{n}_{XYZ}\right)\right]ds_P \quad (3.38)$$

where J_s is the surface current, A is the area of the facet, $\mathfrak{R}_{\text{imp}}$ is the impedance of free space, R is the range from the radar to the origin of the body-fixed coordinates, and $k = 2\pi/\lambda$. Thus, the scattered field from the facet can be calculated by integrating the surface current over the area of the facet. Therefore, the RCS of the facet as a function of R, θ, and φ is obtained. References [12–14] provided MATLAB codes for calculating the RCS of a triangular facet.

The same procedure can be applied to a collection of facets of an object. Thus, the total RCS of the object is a superposition of the RCS contributions of all of the facets.

3.2.4 Radar Backscattering from Rotor Blades

The rotation of a rotor blade can be easily obtained without using ordinary differential equations. The time-varying location and orientation is calculated using a rotation matrix with zero roll and pitch angles. The variation of its yaw angle is determined by the rotation rate and a given initial angle. Based on the location and orientation of a blade, the RCS and the reflected radar signal from the blade can be calculated. Radar reflected signals from all blades can be obtained by the coherent superposition of the reflected signals from each individual blade.

If a coherent radar system transmits a sequence of narrow rectangular pulses with the pulse width Δ and the pulse repetition interval ΔT, the baseband signal in the receiver is

$$s_B(t) = \sum_{k=1}^{n_p}\sum_{n=1}^{N_B}\sqrt{\sigma_n(t)} \cdot rect\left\{t - k \cdot \Delta T - \frac{2R_n(t)}{c}\right\} \cdot \exp\left\{-j2\pi f_c \frac{2R_n(t)}{c}\right\}$$

$$(3.39)$$

where N_B is the total number of blades, n_p is the total number of pulses received during the observation interval, f_c is the radar transmitted frequency, $R_n(t)$ is the distance between the radar and the nth blade at time t, $\sigma n(t)$ is the RCS of the nth rotor blade at time t, and *rect* is the rectangular function defined by $rect(t) = 1$ $(0 < t \leq \Delta)$.

In (3.39), two variables, $\sigma_n(t)$ and $R_n(t)$, must be calculated. The RCS of a perfectly conducting rectangular flat plate, given by (3.37), can be used for the calculation of $\sigma_n(t)$. Equation (3.37) is simple, but only accurate for aspect angles $\theta \leq 20°$. The distance variable $R_n(t)$ is defined from the radar to a scatterer center, such as the centroid, of the rectangular plate. If the rectangular flat plate is relatively large, its centroid can be far away from the tip of the blade. Therefore, if the centroid is assigned as the scattering center, the radar received signal calculated by (3.37) does not include returned signal from the tip of blade. In cases of relatively large rectangular plate, the scatterer center may be assigned directly to the tip of rectangular blade.

The geometry of the radar and rotor blades is shown in Figure 3.16, where the radar is located at $(X_1 = 500m, Y_1 = 0m, Z_1 = 500m)$ with a wavelength of 0.06m at the C-band, the rotor center is located at $(X_0 = 0m, Y_0 = 0m, Z_0 = 0m)$, the length of the blade is $L = 6m$ with its root $L_1 = 0.5m$ and its tip $L_2 = 6.5m$, the width of the blade is $W = 1m$, and the rotation rate is $\Omega = 4$ r/s. The azimuth angle φ and aspect angle θ can be calculated from the

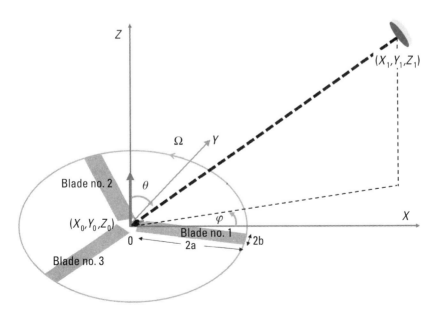

Figure 3.16   Geometry of a radar and rotating rotor blades.

radar location, the rotor location, and the blade geometry. By assigning the scatterer center of each blade to the tip of the blade, the baseband signal in the radar receiver is obtained from (3.39) and the RCS is calculated by (3.37) just for simplicity, even if it is not perfectly accurate. The accuracy of the RCS calculation only determines the magnitude's distribution in the micro-Doppler signature; it does not affect the shape of the signature. After rearranging the n_p range profiles, 2-D pulse-range profiles are shown in Figure 3.17(a) for a two-blade rotor, and its micro-Doppler signature is shown in Figure 3.17(b). The MATLAB code for calculating radar backscattering from rotor blades is provided in this book.

From the range profiles, the rotating blades can be seen around range cell no. 1412 or about a 700-m distance from the radar. However, flashes cannot be seen because of the RCS is only assigned to one scatterer center. To see the flashes in the simulation, a more accurate RCS model is needed.

A simple but more accurate model for calculating radar backscattering from rotating rotor blades is the POFACET model. A rectangular blade is represented by the arrays of triangular facets as shown in Figure 3.18. The scatterer center of each triangle is assumed to be the geometric centroid of its triangle vertices. With the POFACET model, the baseband signal in the radar receiver is modified as

$$s_B(t) = \sum_{k=1}^{n_p} \sum_{n=1}^{N_B} \sum_{m=1}^{N_F} \sqrt{\sigma_{n,m}(t)}\, rect\left\{ t - k\Delta T - \frac{2R_{n,m}(t)}{c} \right\} \exp\left\{ -j2\pi f_c \frac{2R_{n,m}(t)}{c} \right\}$$

$$(3.40)$$

where N_B is the number of blades, N_F is the total number of facets in each blade, n_p is the total number of pulses during the radar observation time interval, and the RCS of each facet $\sigma_{n,m}(t)$ is calculated by using source codes provided in the POFACET [12–14].

Based on the geometry of a rotating three-blade rotor illustrated in Figure 3.18 and the same parameters used in the rotating two-blade rotor, the POFACET model-based radar range profiles and the micro-Doppler signature of the three-blade rotor are shown in Figure 3.19, where the flashes of the rotating blades are seen clearly.

3.2.5 Micro-Doppler Signatures of Rotor Blades

The micro-Doppler signature of the rotating three-blade's rotor, shown in Figure 3.19(b), is obtained from a summation of the range profiles within a range gate where the blades are located and shown in Figure 3.19(a). The

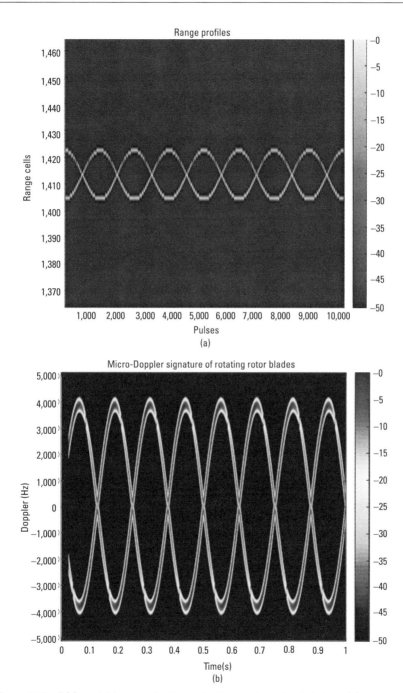

Figure 3.17   RCS model for a perfectly conducting rectangular flat plane: (a) range profiles of a rotating two-blade rotor, and (b) the micro-Doppler signature.

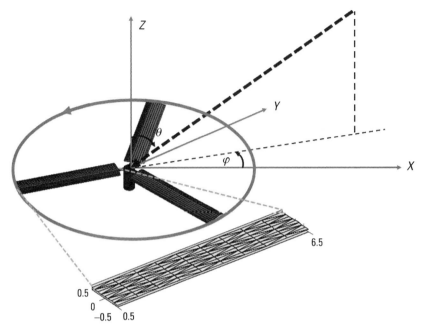

Figure 3.18   A rectangular blade represented by arrays of triangular facets.

joint time-frequency transform used to generate the signature is the STFT. Similarly, the range profiles and the micro-Doppler signature of a rotating two-blade rotor are shown in Figure 3.20, where, compared to the signature of the odd number of blades, different features for an even number of blades can be seen. With the POFACET model of rotor blades, the flashes can be seen in the micro-Doppler signatures. For the two-blade rotor shown in Figure 3.20 with a 4-r/s rotation rate, each blade has 8 flashes in 1.0 second, and the interval between flashes is 0.125 second. For the three-blade rotor in Figure 3.19, there are total of 24 flashes in 1.0 second, and the interval between two successive flashes is 0.0417 second.

Compared to the POFACET model prediction, Figure 3.21 shows the micro-Doppler signature of a two-blade rotor on a scale model helicopter measured by X-band FMCW radar with a wavelength of $\lambda = 0.03$m. The rotation rate of the rotor is about $\Omega = 2.33$ r/s and the blade length is $L = 0.2$m. Thus, the tip velocity is $V_{tip} = 2\pi L\Omega = 2.93$ m/s and the maximum Doppler shift is $\{f_D\}_{max} = 195$ Hz as shown in Figure 3.21. The signature is similar to that of the POFACET model prediction with flashes. From the micro-Doppler

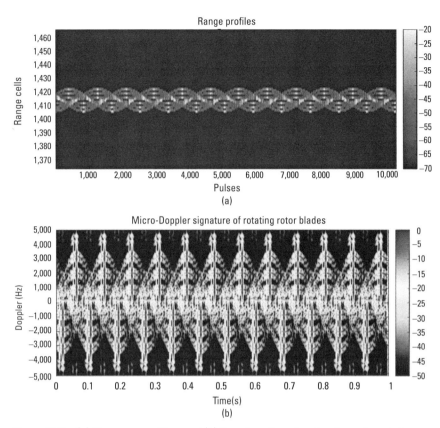

Figure 3.19 (a) The range profiles, and (b) the micro-Doppler signature of a rotating three-blade rotor.

signature, the number of blades, the length of the blade, and the rotation rate of the rotor can be estimated.

3.2.6 Required Minimum PRF

For a pulsed Doppler radar, its pulse repetition frequency (PRF) determines the sampling rate. The required minimum sampling rate must satisfy the Nyquist rate to avoid frequency aliasing. For actual helicopters, the range of their blade tip speed is around 200–230 m/s. With an X-band radar, a blade tip speed of $V_{\text{tip}} = 230$ m/s can generate $\{f_D\}_{\max} = 15$ kHz Doppler shift. Therefore, the required minimum sampling rate is $2 \times \{f_D\}_{\max} = 30$ kHz for a hovering helicopter. If the helicopter has a translational motion with a radial velocity

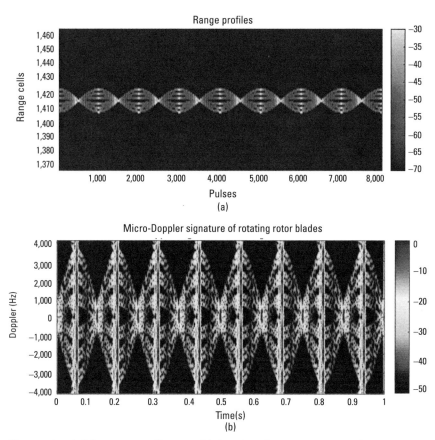

Figure 3.20 (a) The range profiles, and (b) the micro-Doppler signature of a rotating two-blade rotor.

of 100 m/s, the maximum Doppler shift of the helicopter becomes 22 kHz, and the required minimum sampling rate is 44 kHz [8].

Figure 3.22 demonstrates the impact of the sampling rate on the micro-Doppler signatures of rotor blades. In the demonstration, the parameters of a two-blade rotor are the same as described before but with a lower rotation rate, $\Omega = 1$ r/s. In this case, the tip velocity of the blade will be $V_{tip} = 40.84$ m/s or the Doppler shift of the tip is $\{f_D\}_{max} = 1.36$ kHz. Figure 3.22(a) is the micro-Doppler signature of the rotor blades under a sampling rate of 512 samples/second, where the periodic motion of blades cannot be seen. Figure 3.22(b) shows the micro-Doppler signature of the same rotor blades with two times higher sampling rate than 512 samples/second, and the periodic motion

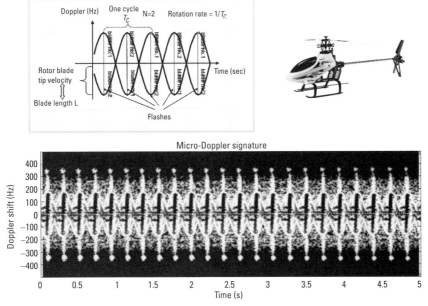

Figure 3.21 The micro-Doppler signature of a rotor with two blades on a model helicopter.

of blades begins to show up, but is incomplete. Figure 3.22(c) is the micro-Doppler signature of the rotor blades with 4 times the 512 sampling rate, and the periodic motion of blades can be seen clearly and is almost completed. In this example, the required minimum sampling rate should be $2 \times \{f_D\}_{max}$ = 2.72 kHz. Therefore, to show the complete micro-Doppler signature, the sampling rate should be higher than 2.72 kHz.

3.2.7 Analysis and Interpretation of the Micro-Doppler Signature of Rotor Blades

Compared to the Doppler spectral signature of helicopter rotors shown in Figure 3.8, the micro-Doppler signature of rotating rotor blades is represented in the joint time-frequency domain to better explore the time-varying Doppler features. The micro-Doppler features of a rotating rotor with two blades and three blades are depicted in Figure 3.23(a, b), respectively. It is obviously that the Doppler patterns of the even number of blades and the odd number of blades are different. Even-number blades generate a symmetric Doppler

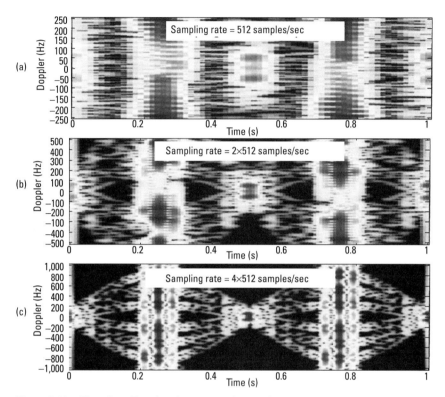

Figure 3.22 The micro-Doppler signatures of a rotating two-blade rotor under different sampling rates: (a) 512 samples/second, (b) 1,024 samples/second, and (c) 2,048 samples/second.

pattern around the mean Doppler frequency, but odd-number blades generate an asymmetric pattern around it. From the micro-Doppler signature of the rotor blades represented in the joint time-frequency domain, the number of blades, the length of blades, the rotation rate of the blades, and the speed of the tip can be estimated. These features are important for identification of an unknown helicopter.

Figure 3.24 is the micro-Doppler signature of a scale model-helicopter using X-band radar. From its symmetric Doppler pattern, the helicopter has two blades. Based on the estimated blade rotation period of $T_C = 0.43$ second and the peak Doppler of $\{f_D\}_{max} = 195$ Hz, the rotation rate Ω, blade diameter $2 \times L$, and tip velocity V_{tip} can be estimated.

Table 3.1 lists a few features of different helicopters. These estimated feature parameters are important for classifying the type of an unknown helicopter.

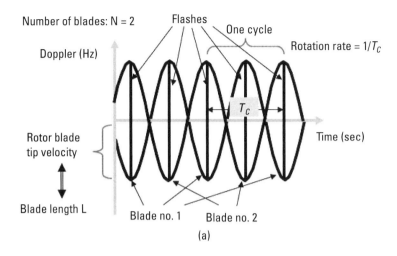

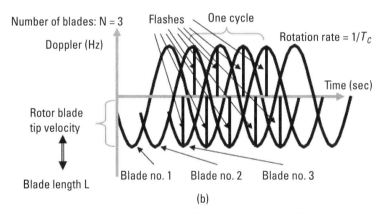

Figure 3.23 The micro-Doppler features of a rotating rotor with (a) two blades and (b) three blades.

3.2.8 Quadrotor and Multirotor Unmanned Aerial Vehicles

Unmanned aerial vehicle (UAV) refers to an aircraft without a human pilot aboard. "Drone" is a popular synonym for the UAV often used for shooting video imagery. Thus, UAV is often used as a term of more advanced unmanned aircraft for which "drone" would not be appropriate.

A rotorcraft with more than two rotors is called a multirotor helicopter, which uses fixed-pitch blades (i.e., their angle of attack is fixed for takeoff,

Micro-Doppler signature of a rotating 2-blade rotor

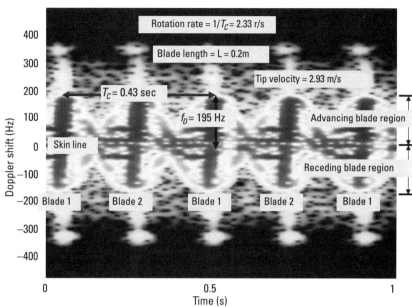

Figure 3.24   The micro-Doppler signature of a scale model helicopter using X-band radar.

Table 3.1
Main Rotor Features of Typical Helicopters

Typical Helicopter	Number of Blades	Diameter (m)	Rotation Rate (r/s)	Tip Velocity (m/s)
AH-1 HUEY COBRA	2	14.63	4.9	227
AH-64 APACHE	4	14.63	4.8	221
UH-60 BLACK HAWK	4	16.36	4.3	221
CH-53 STALLION	7	24.08	2.9	223
MD 500E DEFENDER	5	8.05	8.2	207
A 109 AGUSTA	4	11.0	6.4	222
AS 332 SUPER PUMA	4	15.6	4.4	217
SA 365 DAUPHIN	4	11.94	5.8	218

climb, and cruise). The motion of the rotorcraft is controlled by changing the relative speed of each individual rotor to adjust the thrust and torque produced by them. Usually, the 4-rotor, 6-rotor, and 8-rotor helicopters are called the quadcopter, hexacopter, and octocopter, respectively. Most drones for shooting video imagery belong to the category class-1 UAV. The class-1 micro UAVs are under 2-kg weight and with small payload. Their flight altitude is less than 90m, velocity is very dynamic up to 10 m/s, and their mission radius is under 5 km.

Because radar has distinct advantages over other sensors, such as all-weather, all light conditions, and diverse acquirable information (range, velocity, angle of arrival, micro-Doppler signature), varieties of methods have been proposed for radar detection, tracking, and classification of drones [15–25].

Compared to other air targets, drones are usually small-sized and flying relatively slow and at a low altitude. The RCS of the main body of a drone is about 0.01 m^2 and its rotor blade is even smaller (about 0.001 m^2) [26]. Other small targets flying relatively slow and at a low altitude are birds. Flying birds may cause false alarms to a radar detecting UAVs.

The Doppler spread of flying bird wing flapping has been used for identifying flying birds almost a half-century ago. Sir Eric Eastwood documented the first observations of birds by radar in his book *Radar Ornithology* in 1967. The locomotion of birds, especially the elevation and depression of their wings, can generate special micro-Doppler signatures. Therefore, micro-Doppler signatures have been used to discriminate different flying objects, distinguish flying birds from UAVs, and other air targets.

3.2.8.1 Modeling of Quadrotor UAVs

The geometry of the radar and a quadrotor UAV is shown in Figure 3.25. The radar is located at the origin of the space-fixed coordinates (X, Y, Z) and the quadrotor UAV is centered at the origin of the body-fixed coordinates on the plane $(x, y, z = 0)$. The four rotors are rotating about the z-axis. Assume that the rotors no. 1 and no. 2 are rotating counterclockwise and the rotors no. 3 and no. 4 are rotating clockwise as indicated in Figure 3.25(a). The reference coordinates (X', Y', Z') are parallel to and translated from the space-fixed coordinates located at the same origin as the body-fixed coordinates. The distance from the radar to the origin of the reference coordinates is R_0. The radar observed azimuth and elevation angles of the origin of the reference coordinates are α and β, respectively.

As discussed in Section 3.2, the quadrotor can be modeled by a set of four rotors, where each rotor has N_B blades. To calculate EM scattering from the $4 \times N_B$ rotating blades, for simplicity, the blade can be modeled as a rigid,

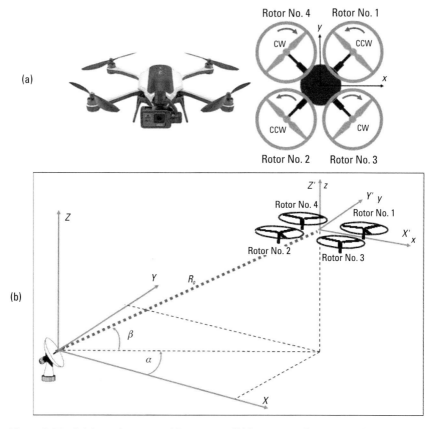

Figure 3.25 (a) A quadcopter and four rotors. (b) Geometry of a radar and the quadrotor.

homogeneous, linear rectangular flat plate. For a perfectly conducting rectangular flat plate, the mathematical formula of its RCS is available in (3.37). The RCS of each blade can be assigned to at the tip of the blade for simplicity.

The radar returned signal from each rotating blade can be calculated by (3.39). The baseband signal returned from the quadcopter is the summation of the returns from the N_R rotors and each rotor has N_B rotating blades:

$$s_\Sigma(t) = \sum_{j=1}^{N_R} \sum_{k=0}^{N_B-1} \sigma_{j,k}(t) \exp\left\{-j\frac{4\pi}{\lambda}R_j(t)\right\} \exp\left\{-j\Phi_k(t)\right\} \qquad (3.41)$$

where the phase function is

$$\Phi_k(t) = \frac{4\pi}{\lambda} L_2 \cos\beta \cos\left(\Omega_j t + \varphi_j + \frac{k2\pi}{N_B}\right), \ \left(j = 1,\ldots N_R; \ k = 0,1,\ldots N_B - 1\right)$$

$$(3.42)$$

and N_B is the total number of blades, N_R is the total number of rotors, $R_j(t)$ is the range from the radar to the center of the jth rotor at time t, $\sigma_{j,k}(t)$ is the RCS of the kth blade on the jth rotor at time t, Ω_j is the rotation rate of the jth rotor, and φ_j is the initial rotating phase of the jth rotor.

3.2.8.2 Micro-Doppler Signatures of Quadrotor UAV

As shown in the geometry of the radar and the quadcopter in Figure 3.25(b), the radar is located at ($X = 0$m, $Y = 0$m, $Z = 0$m) and the center of the quadcopter is located at ($X = 50$m, $Y = 0$m, $Z = 20$m). The center of the quadcopter is at the origin of the body-fixed coordinates and on the plane ($x, y, z = 0$). The four rotors are centered at ($x_1 = 0.2$m, $y_1 = 0.2$m), ($x_2 = -0.2$m, $y_2 = -0.2$m), ($x_3 = 0.2$m, $y_3 = -0.2$m), and ($x_4 = -0.2$m, $y_4 = 0.2$m), respectively. The rotation rate of the rotors is about 100 r/s when hovering and about 150 r/s when at full power.

The power spectrum of the baseband signal is shown in Figure 3.26 and the micro-Doppler signature of the rotors is calculated based on (3.41) and shown in Figure 3.27. Assume that the radar is operating at C-band with a wavelength of 0.0517m. The rotation rate of each rotor is assumed to be 150 r/s, the length of each blade is 7.0 cm with its root at 0.0 cm and its tip at 7.0 cm, and its width is 2.5 cm. Thus, the tip velocity V_{tip} is $2\pi L\Omega = 66$ m/s and the maximum Doppler shift $\{f_D\}_{max}$ is 2.4 kHz. From the micro-Doppler signature, the rotation rate of 150 r/s can be estimated from the periodic pattern as indicated in the figure. The pattern in each period is generated from the four rotors with random initial rotation angles. Thus, the periodic pattern may vary, but the period is closely related to the rotation rate. The MATLAB code for calculating radar backscattering and micro-Doppler signature from a quadcopter is provided in this book.

Some unique micro-Doppler features may be useful for detection and identification of multicopter UAVs. The radial velocity induced micro-Doppler signature is the most useful feature for the multirotor UAVs. However, for some geometry configurations between radar and UAVs, the radial velocity induced micro-Doppler signatures can be inconspicuous. Thus, as introduced in Chapter 1, Section 1.10, the angular velocity induced micro-Doppler modulation is a good complementary feature to the radial induced one. Especially for small UAVs, the angular micro-Doppler signature is always noticeable.

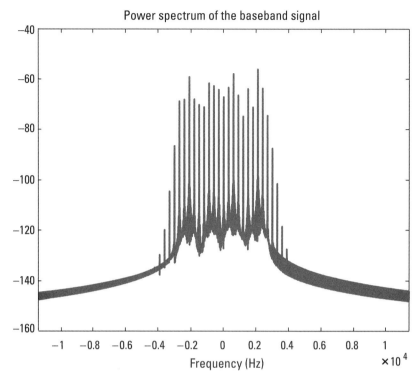

Figure 3.26   Power spectrum of a baseband signal returned from a quadcopter.

3.3   Spinning Symmetric Top

A top stands steadily on a fixed tip point on its symmetric axis and quickly spins about the axis. If the spin axis is inclined, it will rotate sweeping out a vertical cone in a 3-D space as illustrated in Figure 3.28. This type of motion is called the torque-induced precession. The angle between the symmetric axis and the vertical axis, called the precession axis angle, usually varies with time, and the symmetric axis is bobbing up and down, known as nutation. In mechanics, nutation refers to irregularities in the precession caused by the torque applied to the top.

The motion dynamics of a spinning top can be solved by Euler's motion differential equations. When a rigid top rotates about an arbitrary axis with Eulerian angles ψ, θ, and φ in the body-fixed coordinates, the changing rate of these angles described by the Eulerian angles' derivatives vector, $\dot{\Theta} = [\dot{\psi}, \dot{\theta}, \dot{\varphi}]^T$, is related to the angular velocity vector $\Omega = [\Omega_1, \Omega_2, \Omega_3]^T$ through a 3-by-3 Euler angle transform matrix [27]

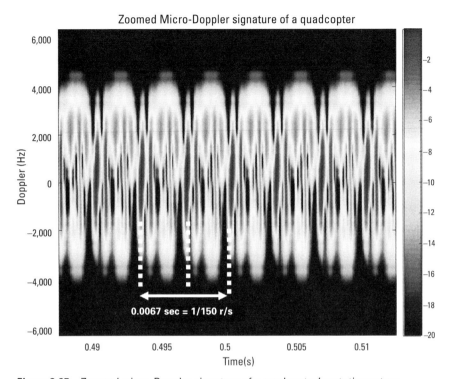

Figure 3.27 Zoomed micro-Doppler signature of a quadcopter's rotating rotors.

$$\mathbf{T} = \begin{bmatrix} \sin\varphi\sin\theta & \cos\varphi & 0 \\ \cos\varphi\sin\theta & -\sin\varphi & 0 \\ \cos\theta & 0 & 1 \end{bmatrix} \tag{3.43}$$

where Ω_1, Ω_2, and Ω_3 are the instantaneous components of the angular velocity with respect to the body-fixed coordinates and T denotes the transposed vector, such that

$$\Omega = \mathbf{T}\dot{\Theta} \tag{3.44}$$

or

$$\begin{bmatrix} \Omega_1 \\ \Omega_2 \\ \Omega_3 \end{bmatrix} = \begin{bmatrix} \sin\varphi\sin\theta\dot{\psi} + \cos\varphi\dot{\theta} \\ \cos\varphi\sin\theta\dot{\psi} - \sin\varphi\dot{\theta} \\ \cos\theta\dot{\psi} + \dot{\varphi} \end{bmatrix} \tag{3.45}$$

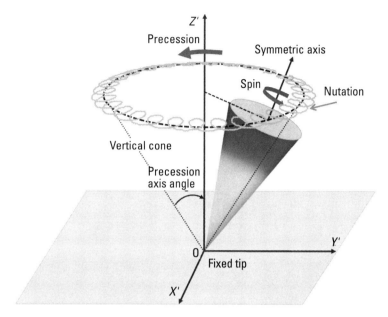

Figure 3.28   Precession of a spinning top.

The inverse Euler angle transform matrix $\mathbf{T}^{-1}$ is [15]

$$\mathbf{T}^{-1} = \begin{bmatrix} \dfrac{\sin\varphi}{\sin\theta} & \dfrac{\cos\varphi}{\sin\theta} & 0 \\[2ex] \cos\varphi & -\sin\varphi & 0 \\[2ex] -\dfrac{\sin\varphi\cos\theta}{\sin\theta} & -\dfrac{\cos\varphi\cos\theta}{\sin\theta} & 1 \end{bmatrix} \tag{3.46}$$

and

$$\dot{\Theta} = \mathbf{T}^{-1}\mathbf{\Omega} \tag{3.47}$$

If an external torque exists, the angular momentum will change, and its changing rate is equal to the torque. For a symmetric top spinning about its symmetric axis and the torque applied about the axis, the angular momentum is $L = I \cdot \mathbf{\Omega}$. The torque $\boldsymbol{\tau}$ is equal to the change rate of the angular momentum:

$$\boldsymbol{\tau} = \frac{d\boldsymbol{L}}{dt} = \boldsymbol{I} \cdot \frac{d\boldsymbol{\Omega}}{dt} \tag{3.48}$$

where $\boldsymbol{\tau} = [\tau_1, \tau_2, \tau_3]^T$ and the inertia tensor $\boldsymbol{I}$ can be a diagonal matrix

$$\boldsymbol{I} = \begin{bmatrix} I_1 & 0 & 0 \\ 0 & I_2 & 0 \\ 0 & 0 & I_3 \end{bmatrix} \tag{3.49}$$

if the principal axes are used as the coordinate axes.

If only the external torque component τ_3 is applied such that the Euler angle φ increases, according to the Lagrangian mechanics [28], the Lagrangian equation is

$$\frac{d}{dt}\left(\frac{\partial E_{\text{Rot}}}{\partial \dot{\varphi}}\right) - \frac{\partial E_{\text{Rot}}}{\partial \varphi} = \tau_3 \tag{3.50}$$

where $E_{\text{Rot}} = \frac{1}{2}(I_1\Omega_1^2 + I_2\Omega_2^2 + I_3\Omega_3^2)$ is the kinetic energy of the rotating top given by (2.33).

Based on (3.45), (3.48) becomes $I_3\dot{\Omega}_3 - (I_1 - I_2)\Omega_1\Omega_2 = \tau_3$, which is the Euler equation for one of the principal axes. The whole Euler equations for principal axes are derived as the differential equations in (2.37):

$$I_1\frac{d\Omega_1}{dt} + \left(I_3 - I_2\right)\Omega_2\Omega_3 = \tau_1$$

$$I_2\frac{d\Omega_2}{dt} + \left(I_1 - I_3\right)\Omega_3\Omega_1 = \tau_2$$

$$I_3\frac{d\Omega_3}{dt} + \left(I_2 - I_1\right)\Omega_1\Omega_2 = \tau_3$$

3.3.1 Force-Free Rotation of a Symmetric Top

For a symmetric top, the principal moments I_1 is equal to I_2. If there is no external torque on the top, the symmetric top will be rotating about an arbitrary axis with an angular velocity vector $\boldsymbol{\Omega} = [\Omega_1, \Omega_2, \Omega_3]^T$, where Ω_1, Ω_2, and Ω_3 are the instantaneous components of its angular velocity with respect to the principal axes. Then the Euler equations become

$$I_1 \frac{d\Omega_1}{dt} + \left(I_3 - I_2\right)\Omega_2\Omega_3 = 0$$

$$I_2 \frac{d\Omega_2}{dt} + \left(I_1 - I_3\right)\Omega_3\Omega_1 = 0 \tag{3.51}$$

$$I_3 \frac{d\Omega_3}{dt} = 0$$

They can be rewritten as

$$\frac{d\Omega_1}{dt} = -\left[\frac{\left(I_3 - I_1\right)}{I_1}\Omega_3\right]\Omega_2$$

$$\frac{d\Omega_2}{dt} = \left[\frac{\left(I_3 - I_1\right)}{I_1}\Omega_3\right]\Omega_1 \tag{3.52}$$

$$\frac{d\Omega_3}{dt} = 0$$

From the third equation, Ω_3 must be a constant: $\Omega_3 = C$. Differentiating the first equation and substituting from the second equation, and differentiating the second equation and substituting from the first one, two simple harmonic motion equations can be derived:

$$\frac{d^2\Omega_1}{dt^2} = -\left[\frac{\left(I_3 - I_1\right)}{I_1}\Omega_3\right]^2 \Omega_1$$

$$\frac{d^2\Omega_2}{dt^2} = -\left[\frac{\left(I_3 - I_1\right)}{I_1}\Omega_3\right]^2 \Omega_2 \tag{3.53}$$

The solutions of these simple harmonic motion equations are

$$\Omega_1 = \Omega_{ini} \cos\left(\Psi t + \Psi_{ini}\right)$$

$$\Omega_2 = \Omega_{ini} \sin\left(\Psi t + \Psi_{ini}\right) \tag{3.54}$$

$$\Omega_3 = C$$

where $\Omega_{ini} = (\Omega_1^2 + \Omega_2^2)^{1/2}$ is the initial amplitude, Ψ_{ini} is the initial phase angle at $t = 0$, and Ψ is the precession angular velocity defined by $\Psi = \Omega_3(I_3 - I_1)/I_1$. Equation (3.54) indicates that a force-free symmetric top will rotate about the principal axes with an angular velocity vector $\mathbf{\Omega} = [\Omega_1, \Omega_2, \Omega_3]^T$, where Ω_3 is a constant and $(\Omega_1^2 + \Omega_2^2)^{1/2}$ is also a constant.

3.3.2 Torque-Induced Rotation of a Symmetric Top

Figure 3.29 illustrates a system model of a spinning symmetric top standing steadily on a fixed tip point of the top. The mass of the top is m and the principal moments of inertia with respect to the fixed-body coordinates are I_1, I_2, and I_3. If the distance from the center of mass to the fixed tip point is L, then, under gravitational force, the Euler differential equations become [27]:

$$\left(I_1 + mL^2\right)\frac{d\Omega_1}{dt} = \left(I_2 - I_3 + mL^2\right)\Omega_2\Omega_3 + mgL\cos\varphi\sin\theta$$

$$\left(I_2 + mL^2\right)\frac{d\Omega_2}{dt} = \left(I_3 - I_1 - mL^2\right)\Omega_1\Omega_3 - mgL\sin\varphi\sin\theta \qquad (3.55)$$

$$I_3\frac{d\Omega_3}{dt} = \left(I_1 - I_2\right)\Omega_1\Omega_2$$

where φ is the spinning angle and θ is the nutation angle. The angle ψ in Figure 3.29 is the precession angle.

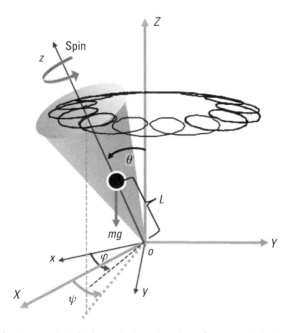

Figure 3.29 System model of a force-induced motion of a symmetric top.

In order to incorporate the top's motion into the EM simulation, the set of Euler equations (3.55) must be solved. Thus, the nonlinear dynamics of the top motion can be obtained. Under the gravity force, the spinning top with a fixed standing point should have a precession motion about an axis. Given the mass of the top $m = 25$ kg, the distance between the center of mass (CM) and the fixed standing point $L = 0.563$m, the moments of inertia $I_1 = I_2 = 0.117$ kg m^2 and $I_3 = 8.5$ kg m^2, the initial angle $\theta_0 = 20°$, the initial spinning velocity $d\varphi_0/dt = 3 \times 2\pi$ rad/sec, the initial precession velocity $d\psi_0/dt = 0.5 \times 2\pi$ rad/sec, and the initial nutation velocity $d\theta_0/dt = 0$, Figure 3.30 shows the angular velocities and the dynamic Euler angles. Figure 3.31 shows the position of the CM and the trajectory of the CM of the top motion.

3.3.3 RCS Model of a Symmetric Top

A symmetric top can be any symmetric geometric shape, such as a cone, a truncated cone (frustum), a cylinder, a sphere, or an ellipsoid. The mathematical

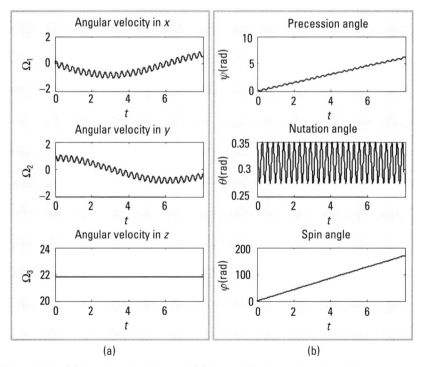

Figure 3.30 (a) Angular velocities, and (b) dynamic Euler angles of a spinning top.

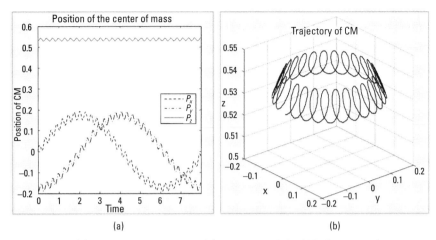

Figure 3.31 (a) Position of the CM, and (b) trajectory of the CM of a top motion.

formulas for calculating the RCS of these simple geometric shapes can be found in [3].

For a truncated cone as illustrated in Figure 3.32, the half cone angle α is determined by $\tan\alpha = (r_2 - r_1)/h$, where h is the height of the truncated cone or frustum. The monostatic RCS of the frustum is [3]

$$RCS_{\text{frustum}} = \begin{cases} \dfrac{8\pi\left(z_2^{3/2} - z_1^{3/2}\right)^2 \sin\alpha}{9\lambda(\cos\alpha)^4} & \text{(at normal incidence)} \\[3mm] \dfrac{\lambda z \tan\alpha}{8\pi\sin\theta}\left[\tan(\theta-\alpha)\right]^2 & \text{(for non-normal incidence)} \end{cases} \tag{3.56}$$

where λ is the wavelength, and z_1, z_2 are indicated in Figure 3.32.

3.3.4 Radar Backscattering from a Symmetric Top

The location and orientation of any point in a rotating symmetric top can be calculated by a rotation matrix and are time-varying. Based on the calculated location and orientation at each time instant, the RCS of the top can be calculated in terms of a RCS model. Then, given radar parameters and a signal waveform, the returned signal from the spinning top can be calculated. If a coherent radar system transmits a sequence of narrow rectangular pulses with a transmitted frequency f_c, a pulse width Δ, and a pulse repetition interval ΔT, the baseband signal in the radar receiver is

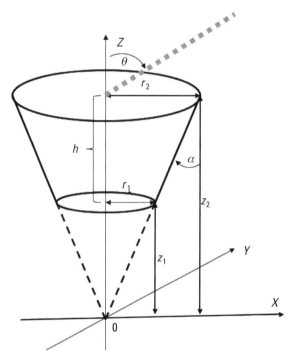

Figure 3.32   The geometry of a frustum.

$$s_B(t) = \sum_{k=1}^{n_p} \sqrt{\sigma(t)}\, rect\left\{ t - k\Delta T - \frac{2R(t)}{c} \right\} \exp\left\{ -j2\pi f_c \frac{2R(t)}{c} \right\} \quad (3.57)$$

where $\sigma(t)$ is the RCS of the top, n_p is the total number of pulses received during the observation time, $R(t)$ is the distance between the radar and the top at time t.

The RCS formula (3.56) of a truncated cone is used for calculating $\sigma(t)$, and the distance $R(t)$ is calculated from the radar to the center of mass of the top.

3.3.5 Micro-Doppler Signatures Generated by a Precessing Top

Given a radar location at $(X = 20\text{m}, Y = 0\text{m}, Z = 0\text{m})$ and the tip of the top at $(X = 0\text{m}, Y = 0\text{m}, Z = 0\text{m})$, the geometric configuration of the radar and the top is illustrated in Figure 3.33.

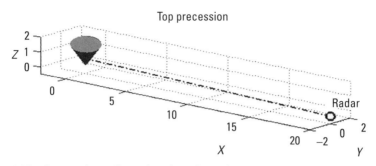

Figure 3.33   Geometric configuration of a radar and top.

After rearranging the n_p range profiles, the 2-D range profiles of the spinning and precessing top are shown in Figure 3.34(a), and the micro-Doppler signature of the top is shown in Figure 3.34(b). In the range profiles, the rotating top can be seen around the range cell no. 667 or about 20m of the distance from the radar. The micro-Doppler signature of the spinning top is obtained from the summation of those range profiles that are within the range gate around 20m of range, where the top is located. The joint time-frequency transform used to generate the signature is the simple STFT. The MATLAB code for calculating radar backscattering from a spinning top is provided in this book.

3.3.6   Analysis and Interpretation of the Micro-Doppler Signature of a Precessing Top

During the radar observation time interval, the simulated spinning and pre-cessing top completed one cycle of precession and 27 cycles of nutation. The Doppler modulation by precession and nutations is shown clearly in the micro-Doppler signature in Figure 3.35. The RCS produced by the upper circular plate is not significant. However, the reflection from the edge of the upper circular plate, which is an interesting feature, is not considered in the simple RCS model.

A more accurate simulation that shows the edge reflection of the upper circular plate can be found in [29, 30], where the mass of the top is $m = 25$ kg, the distance between the center of mass and the fixed tip point is $L = 0.563$m, the moments of inertia $I_1 = I_2 = 0.117$ kg m^2 and $I_3 = 8.5$ kg m^2, and the initial nutation angle $\theta_0 = 20°$. The radar is an X-band radar with a transmitted frequency of 10 GHz and a bandwidth of 500 MHz, located at

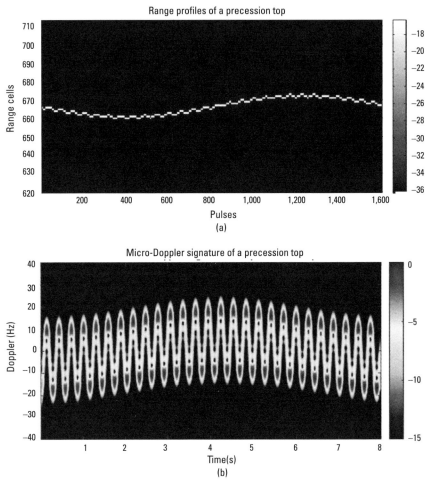

Figure 3.34 (a) Range profiles of a spinning and precessing top, and (b) the micro-Doppler signature of the top.

a distance of 12m from the tip of the top. The more accurate RCS prediction model utilizes a multitude of different backscattering algorithms, including geometrical optics, physical optics, the physical theory of diffraction, and the method of moments solutions.

The micro-Doppler signature of the above spinning top is shown in Figure 3.36. From the signature, about one cycle of the precession during the 5.3-second observation time duration can be seen clearly. There are also 12.5 cycles of nutations as shown in Figure 3.36, and the Doppler modulation produced by the top upper circular plate disk is marked in Figure 3.36.

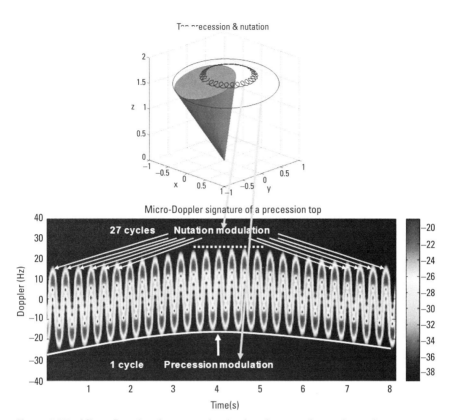

Figure 3.35   Micro-Doppler signature of a simulated precessing and nutating top.

It should be emphasized that, in a precessing top, the inertia ratio I_1/I_3 is an important characteristic. It can be estimated from the measured precession angular velocity, the spinning angular velocity, and the precession angle.

3.4   Micro-Doppler Signatures of Reentry Vehicles

A reentry vehicle (RV) means the part of a ballistic missile that carries a warhead or a decoy that re-enters to the Earth. After being released in space, the RV must spin to keep its orientation. Under the force of gravity, the spinning RV will have precession motion about an axis, N, and bobbing up and down, known as the nutation. Since RVs and decoys have different micromotions, the micro-Doppler analysis introduced in this book can be applied to classify, recognize, and identify ballistic targets. Thus, the study of micromotion

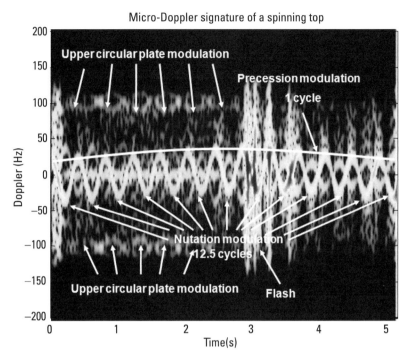

Figure 3.36   The micro-Doppler signature produced by a spinning, precessing, and nutating top with an upper circular plate disk (*After:* [29].)

features and extraction of micro-Doppler signatures from ballistic targets become important issues [31–37].

Based on the discussion in Section 3.3 on spinning symmetric top, the spinning RV becomes a natural extension of research on a spinning symmetric top.

Figure 3.37 illustrates a spinning cone-shaped RV with precession and nutation. The cone is spinning around the symmetric axis. The spinning axis is also rotating about a coning axis, N, that intersects with the spinning axis at a point S. If the spinning axis does not remain at a constant angle with the coning axis, it must oscillate up and down, called the nutation. As indicated in the figure, φ is the angle of spinning rotation called the spin angle, ψ is the angle of precessing rotation called the precession angle (it is different from the precession axis angle defined in Section 3.3), and θ is the angle of oscillation called the nutation angle.

Figure 3.38 is the geometry of a radar and a cone-shaped RV with spinning, precession and nutation. The RV with coning motion along the axis

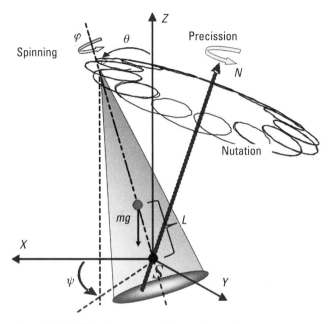

Figure 3.37 A spinning RV also has precession and nutation.

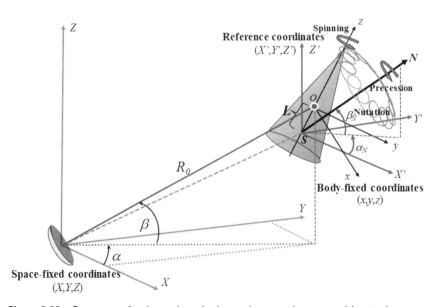

Figure 3.38 Geometry of radar and a spinning and precessing cone with nutation.

is SN. The reference coordinate system (X', Y', Z') is parallel to the radar coordinates (X, Y, Z) and its origin is located at the point S. Assuming that the azimuth and elevation angles of the RV mass center O with respect to the radar are α and β, respectively, then the azimuth and elevation angles of the precession axis SN with respect to the reference coordinates (X', Y', Z') are α_N and β_N, respectively. The distance between the radar and the mass center of the RV is R_0 and the distance from the mass center O to the origin of the reference coordinates S is L.

3.4.1　Mathematical Model of a Cone-Shaped RV

The spinning matrix and coning matrix have been derived in Chapter 2. The precession matrix is derived from the spinning matrix, coning matrix, and nutation matrix. According to the Rodrigues rotation formula, given the unit vector of the spinning axis $\boldsymbol{u}_{spin} = (u_{spin,x}, u_{spin,y}, u_{spin,z})^T$, the vector of angular velocity of spinning $\boldsymbol{\omega}_{spin} = (\omega_{spin,x}, \omega_{spin,y}, \omega_{spin,z})^T$, or the scalar angular velocity $\Omega_{spin} = \|\boldsymbol{\omega}_{spin}\|$, and the unit vector of spinning angular velocity $\boldsymbol{\omega}'_{spin} = (\omega'_{spin,x}, \omega'_{spin,y}, \omega'_{spin,z})^T$, the spinning matrix is

$$\Re_{spin}(t) = I + \hat{\omega}'_{spin} \sin\left(\Omega_{spin} t\right) + \left(\hat{\omega}'_{spin}\right)^2 \left[1 - \cos\left(\Omega_{spin} t\right)\right] \quad (3.58)$$

where $\hat{\omega}'_{spin}$ is a skew symmetric matrix

$$\hat{\omega}'_{spin} = \begin{bmatrix} 0 & -\omega'_{spin,z} & \omega'_{spin,y} \\ \omega'_{spin,z} & 0 & -\omega'_{spin,x} \\ -\omega'_{spin,y} & \omega'_{spin,x} & 0 \end{bmatrix}$$

Similarly, the coning matrix is

$$\Re_{coning}(t) = I + \hat{\omega}'_{coning} \sin\left(\Omega_{coning} t\right) + \left(\hat{\omega}'_{coning}\right)^2 \left[1 - \cos\left(\Omega_{coning} t\right)\right] \quad (3.59)$$

where $\hat{\omega}'_{coning}$ is a skew symmetric matrix

$$\hat{\omega}'_{coning} = \begin{bmatrix} 0 & -\sin\beta_N & \sin\alpha_N \cos\beta_N \\ \sin\beta_N & 0 & -\cos\alpha_N \cos\beta_N \\ -\sin\alpha_N \cos\beta_N & \cos\alpha_N \cos\beta_N & 0 \end{bmatrix}$$

Since nutation is an oscillation, the nutation matrix is expressed as

$$
\mathfrak{R}_{\text{nut}}(t) = \left[u_{\text{nut},x}, u_{\text{nut},y}, u_{\text{nut},z} \right] \cdot \begin{bmatrix} \cos\theta(t) & -\sin\theta(t) & 0 \\ \sin\theta(t) & \cos\theta(t) & 0 \\ 0 & 0 & 1 \end{bmatrix} \cdot \begin{bmatrix} u_{\text{nut},x} \\ u_{\text{nut},y} \\ u_{\text{nut},z} \end{bmatrix} \tag{3.60}
$$

where $\theta(t)$ is the nutation angle and $\boldsymbol{u}_{\text{nut}}$ is the unit nutation vector.

Thus, the combination of spinning, coning and nutation becomes the micromotion matrix of the RV

$$
\mathfrak{R}_{RV} = \mathfrak{R}_{\text{spin}} \mathfrak{R}_{\text{coning}} \mathfrak{R}_{\text{nut}} \tag{3.61}
$$

Then, the micro-Doppler modulation of precession motion with spinning, coning, and nutation is given by

$$
f_{mD}(t)\big|_{S+C+N} = \frac{2}{\lambda} \left[\begin{array}{c} \left(\dfrac{d}{dt}\mathfrak{R}_{\text{coning}} \right)\mathfrak{R}_{\text{spin}}\mathfrak{R}_{\text{nut}} + \mathfrak{R}_{\text{coning}}\left(\dfrac{d}{dt}\mathfrak{R}_{\text{spin}} \right)\mathfrak{R}_{\text{nut}} \\[2ex] + \mathfrak{R}_{\text{coning}}\mathfrak{R}_{\text{spin}}\left(\dfrac{d}{dt}\mathfrak{R}_{\text{nut}} \right) \end{array} \right]_{\text{radial}} \tag{3.62}
$$

3.4.2 Motion Dynamic Model of a Cone-Shaped RV

The motion dynamics of the cone-shaped RV are the same as those of the torque-induced rotation of a symmetric top.

When a rigid body rotates about an arbitrary axis with Eulerian (spin, precession, and nutation) angles in the body-fixed coordinates, the changing rate of these angles (i.e., the Eulerian angles derivatives vector) $\dot{\Theta} = [\dot{\psi}, \dot{\theta}, \dot{\varphi}]^T$ is related to the angular velocity vector $\Omega = [\Omega_1, \Omega_2, \Omega_3]^T$, where Ω_1, Ω_2, and Ω_3 are the instantaneous components of its angular velocity with respect to the body-fixed coordinates. The relationship between Ω and $\dot{\Theta}$ is through an operator T of the Euler angle rotation:

$$
T(\psi, \theta, \varphi) = \begin{bmatrix} \sin\varphi\sin\theta & \cos\varphi & 0 \\ \cos\varphi\sin\theta & -\sin\varphi & 0 \\ \cos\theta & 0 & 1 \end{bmatrix} \tag{3.63}
$$

such that

$$\Omega = T(\psi,\theta,\varphi)\dot{\Theta} \tag{3.64}$$

Let the mass of the RV be m and the principal moments of inertia with respect to the body-fixed coordinates be I_1, I_2, and I_3. Because the distance from the center of mass to the origin of the reference coordinates is L, then under gravitational force, the Euler differential equations are

$$\left(I_1 + mL^2\right)\frac{d\Omega_1}{dt} = \left(I_2 - I_3 + mL^2\right)\Omega_2\Omega_3 + mgL\cos\varphi\sin\theta$$

$$\left(I_2 + mL^2\right)\frac{d\Omega_2}{dt} = \left(I_3 - I_1 + mL^2\right)\Omega_1\Omega_3 + mgL\sin\varphi\sin\theta \tag{3.65}$$

$$I_3\frac{d\Omega_3}{dt} = \left(I_1 - I_2\right)\Omega_1\Omega_2$$

where φ is the spinning angle, θ is the nutation angle, and ψ is the precession angle as indicated in Figure 3.37.

Assuming the mass $m = 25$ kg, the distance $L = 0.5$m, the moments of inertia $I_1 = I_2 = 0.117$ kg m^2 and $I_3 = 8.5$ kg m^2, and the initial nutation angle $\theta_0 = 20°$, the initial spinning velocity $d\varphi_0/dt = 20$ rad/sec, the initial precession velocity $d\psi_0/dt = -6$ rad/sec, and the initial nutation velocity $d\theta_0/dt = 0$, by solving the set of motion differential equations, the nonlinear dynamics of the RV can be obtained as shown in Figure 3.39.

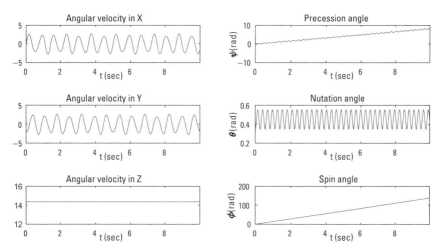

Figure 3.39 Angular velocities and Euler angles of precession, nutation, and spin.

3.4.3 Micro-Doppler Signature Analysis

A coherent radar system transmits a sequence of narrow rectangular pulses with a transmitted frequency f_c, a pulse width Δ, and a pulse repetition interval ΔT, the baseband signal in the radar receiver is

$$s_B(t) = \sum_{k=1}^{n_p} \sqrt{\sigma(t)}\, rect\left\{ t - k\Delta T - \frac{2R(t)}{c} \right\} \exp\left\{ -j2\pi f_c \frac{2R(t)}{c} \right\} \quad (3.66)$$

where $\sigma(t)$ is the RCS of the RV, n_p is the total number of pulses received during the observation time, $R(t)$ is the distance between the radar and the mass center of the RV at time t. The RV can be modeled as a frustum cone shape.

The radar is an X-band radar at 10-GHz transmit frequency with 500-MHz bandwidth, located at 20-m distance from the center of mass of the RV. During a 10-second observation time, the RV completed 1.34 cycles of the precession angle and has 37 cycles of nutation oscillation. The Doppler modulation by precession and nutation is clearly shown in the micro-Doppler signature in Figure 3.40.

3.4.4 Summary

From the micro-Doppler signatures of RVs, some important micromotion parameters, such as spin rate, precession rate, nutation angle, and inertia ratio,

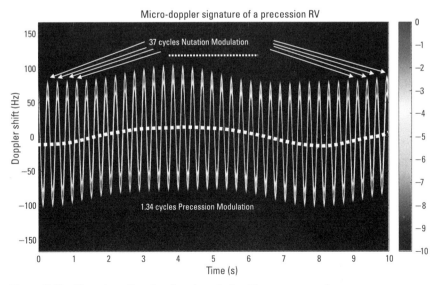

Figure 3.40 The micro-Doppler signature derived from a precession cone.

can be estimated. The precession and nutation of a ballistic missile warhead and the wobble motion of a decoy are two typical micromotions. Their different micro-Doppler signatures may be used to identify a warhead from decoys. It was found that since the inertial parameters of an object are closely related to the state of its micromotion, the inertial ratio of a rigid cone can serve as an important merit index of the object for the target discrimination.

3.5   Wind Turbines

Since the use of wind energy has been dramatically increasing, the large numbers of wind turbines and the large RCS of the wind turbine blade become challenges to current radar systems as illustrated in Figure 3.41. A typical turbine could have a RCS of the order of 60 dBsm (or 10^6 m^2) at X-band [38].

The massive construction of wind turbines with large size and high linear velocity become a new form of radar clutter-like interferences that degrade radar functionalities, cause spurious detection and tracking of targets, and disturb estimated target parameters. The impacts of wind turbines on radar performance, including air traffic control systems, navigation systems, weather radar systems, and other primary or secondary radar systems, have been investigated and reported [38–48].

The Doppler frequency shift produced by the rotating blades of wind turbines has an impact on the ability of radars to discriminate the wind turbine from a flying aircraft. Even if the rotor rotation rate is low, however, a large blade diameter gives a tip velocity falling in the range of 50 to 150 m/s,

Figure 3.41   Wind turbines observed by radar systems.

which is within a speed range of an aircraft. The large physical size of the blade produces a substantial RCS and a broader spectrum. Consequently, the wind turbine blades viewed from radars appear as a moving aircraft.

Although the development of new algorithms to suppress a wind turbine's clutter is an important task and challenge, it is out of the scope of this book.

3.5.1 Micro-Doppler Signatures of Wind Turbines

A wind turbine normally consists of a tower, a power-generating nacelle, hub, and turbine blades. The first three parts provide the major contributions to the radar returns. Because their Doppler frequency components fall into near Doppler-zero frequency in the frequency domain, they can be easily suppressed by conventional notch filters. The power-generating nacelle slowly rotates its direction to enable turbine blades to face to the wind. Even if it is slow rotating, the nacelle can still be considered a virtually stationary object.

The actual moving parts of the wind turbine are the turbine blades. The blade is a large, aerodynamically shaped structure that operates like a rotor blade of a helicopter. Its motion kinematic and dynamic properties are similar to those of helicopter rotor blades. Thus, the mathematical model, motion dynamics, and EM scattering model of helicopter rotor blades are also suitable for wind turbines.

In addition to the rotating blades, the wind turbine also has the other two degrees of freedom (i.e., yaw and pitch). The yaw is made according to the wind direction to achieve maximum efficiency. The pitch is made for changing the angle of attack to adapt the blade rotation rate to wind speed. The blade rotation rate usually varies from 10 to 20 rpm, which makes rotating-types of the micro-Doppler signature. The yaw and pitch are slow motions that lead to different micro-Doppler spectrum patterns and make the wind turbine's unique EM signatures. The wind turbine features with respect to the wind turbine's shape, material, pitch/yaw positions, and rotation of blades make radar micro-Doppler signatures more specific and complicated.

An example of micro-Doppler signature of a real-world wind turbine was given in [48]. The field study used an experimental X-band polarized weather radar and the wind turbine is a state-of-the-art GE 1.6-MW. The wind turbine was at about 7,350-m distance from the radar and the rotor blades were nearly parallel to the radar LOS at a 270° aspect. The measured micro-Doppler signature of the wind turbine is shown in Figure 3.42, where the Doppler frequency axis is converted to the radial velocity axis.

The GE 1.6-MW turbine has variable rotation rate. From the micro-Doppler signature, the estimated tip speed is found to be close to 80 m/s. The blade rotating period is about 3.4 seconds or the rotation rate is 17.6 rpm. Because the actual rotor diameter is 82.5m, the actual tip speed should be 76 m/s, which is near the observed tip speed from the micro-Doppler signature.

3.5.2 Analysis and Interpretation of the Micro-Doppler Signature of Wind Turbines

In a similar way to the micro-Doppler signatures of helicopter rotors in Section 3.2.5, the micro-Doppler signature of wind turbines has strong components near zero Doppler due to strong stationary reflections from the tower, nacelle, and other ground clutter. Its RCS is much higher than the RCS of the helicopter's rotor blades. However, the oscillation rate is much lower than the helicopter rotors. It is easy to distinguish the micro-Doppler signatures of wind turbines from that of helicopters, especially when a helicopter is moving and the center line of its micro-Doppler signature is shifted from zero Doppler.

The micro-Doppler signature of real-world wind turbine in Figure 3.42 shows some interesting features. The positive and negative velocity sides of the signature are quite different from each other. The negative side is the blade rotating toward the radar and has obviously lower power than the positive side. At the 270° aspect, the negative flashes correspond to the leading edge sweeping downward and coming toward radar, while the positive flashes are the trailing edge sweeping upward and going away from the radar. Because the leading edge is thicker, the flash from leading edge is stronger than that from trailing edge.

In addition, micro-Doppler signatures of wind turbines may also have Doppler components due to multiple bounces. The multibounce effect may occur if radar waves are reflected off two different surfaces before returning to the radar receiver. In wind turbines, the multibounce occurs while radar transmitted waves are reflected from large turbine blades to the turbine tower and then again to the blades before returning to the radar receiver.

Compared with a single turbine, the micro-Doppler signatures from multiple wind turbines are much more complicated. All turbines in a wind farm may not be aligned to the same direction and turbine directions may vary widely. The effects of multiple turbines on the RCS are beyond the scope of this book.

3.5.3 Simulation Study on Wind Turbines

In a simulation of wind turbines, the RCS of a wind turbine's blades can be estimated by RCS prediction methods. However, the large-sized blades

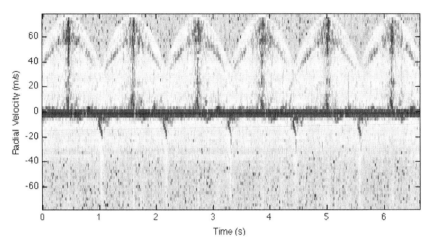

Figure 3.42 The micro-Doppler signature of a real-world wind turbine GE 1.6-MW. (*After:* [48].)

and multiple degrees of freedom of wind turbines make an extremely huge computational load. Thus, a simple RCS prediction should be used for the wind turbine.

A RCS prediction code based on the PO method called the POFACET in MATLAB is available [49]. The PO method is a high-frequency approximation to estimate the surface current induced on a body. In using the POFACET method, an object is approximated by large number of triangular meshes, called facets that produce a continuous surface of the object. The total RCS of the object is the superposition of the square root of the magnitude of each individual facet's RCS. The scattered field of each triangle is computed by assuming that the triangle is isolated and other triangles are not present. Shadowing is only considered by a facet to be completely illuminated or completely shadowed by the incident wave.

Using the POFACET method, the micro-Doppler signature of a rotating three-blade turbine rotor is shown in Figure 3.43, where the length of the blade is 20m and the width of the blade is 1m, and the rotation rate is 0.25 r/s or rotating one cycle in 4 seconds. The radar is operating at C-band with a frequency of 5.0 GHz. In the micro-Doppler signature of the wind turbine blades, flashes at the receding and approaching points can be seen. There are total of eight flashes from three blades in 5 seconds. The interval between two flashes produced by two successive blades is about 1.33 seconds. MATLAB codes similar to calculating helicopter rotor blades are provided in this book. These functions that are related to the POFACET originated from [49].

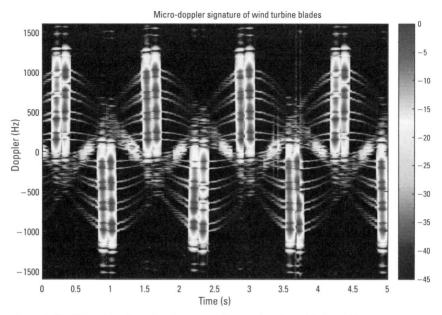

Figure 3.43   The micro-Doppler signature of a rotating three-blade turbine rotor.

References

[1]   Knott, E. F., J. F. Schaffer, and M. T. Tuley, *Radar Cross Section*, 2nd ed., Norwood, MA: Artech House, 1993.

[2]   Shirman, Y. D., (ed.), *Computer Simulation of Aerial Target Radar Scattering, Recognition, Detection, and Tracking*, Norwood, MA: Artech House, 2002.

[3]   Mahafza, B., *Radar Systems Analysis and Design Using MATLAB*, 3rd ed., Chapman & Hall/CRC, 2013.

[4]   Youssef, N., "Radar Cross Section of Complex Targets," *Proc. of IEEE*, Vol. 77, No. 5, 1989, pp. 722–734.

[5]   MacKenzie, J. D., et al., "The Measurement of Radar Cross Section," *Proceedings of the Military Microwaves '86 Conference*, June 24–26, 1986, pp. 493–500.

[6]   Shi, N. K., and F. Williams, "Radar Detection and Classification of Helicopters," U.S. Patent No. 5,689,268, November 18, 1997.

[7]   Chen, V. C., "Radar Signatures of Rotor Blades," *Proceedings of SPIE on Wavelet Applications* VIII, Vol. 4391, 2001, pp. 63–70.

[8]   Martin, J., and B. Mulgrew, "Analysis of the Theoretical Radar Return Signal from Aircraft Propeller Blades," *IEEE 1990 International Radar Conference*, 1990, pp. 569–572.

[9] Misiurewicz, J., K. Kulpa and Z. Czekala, "Analysis of Recorded Helicopter Echo," *IEE Radar 97, Proceedings*, 1997, pp. 449–453.

[10] Pouliguen, P., et al., "Calculation and Analysis of Electromagnetic Scattering by Helicopter Rotating Blades," *IEEE Transactions on Antennas and Propagation*, Vol. 50, No. 10, October 2002, pp. 1396–1408.

[11] Anderson, W.C., *The Radar Cross Section of Perfectly Conducting Rectangular Flat Plates and Rectangular Cylinders: A Comparison of Physical Optics, GTD and UTD Solutions*, Technical report ERL-0344-TR DSTO, Australia, 1985.

[12] Chatzigeorgiadis, F., "Development of Code for Physical Optics Radar Cross Section Prediction and Analysis Application," Master's Thesis, Naval Postgraduate School, Monterey, CA, September 2004.

[13] Chatzigeorgiadis, F., and D. Jenn, "A MATLAB Physical-Optics RCS Prediction Code," *IEEE Antenna and Propagation Magazine*, Vol. 46, No. 4, 2004, pp. 137–139.

[14] Garrido, E. E., "Graphical User Interface for Physical Optics Radar Cross Section Prediction Code," Master's Thesis, Naval Postgraduate School, Monterey, CA, September 2000.

[15] Singh, A. K., and Y. -H. Kim, "Automatic Measurement of Blade Length and Rotation Rate of Drone Using W-Band Micro-Doppler Radar," *IEEE Sensors Journal*, Vol. 18, No. 5, 2018, pp. 1895–1902.

[16] Rahman, S., and D. Robertson, "Time-Frequency Analysis of Millimeter-Wave Radar Micro-Doppler Data from Small UAVs," *2017 Sensor Signal Processing for Defense Conference*, 2017, pp. 1–5.

[17] Fuhrmann, L., et al., "Micro-Doppler Analysis and Classification of UAVs at Ka Band," *2017 18th International Radar Symposium (IRS)*, 2017.

[18] Molchanov, P., et al., "Classification of Small UAVs and Birds by Micro-Doppler Signatures," *International Journal of Microwave and Wireless Technologies*, Vol. 6, No. 3-4, 2014, pp. 435–444.

[19] Ritchie, M., et al., "Monostatic and Bistatic Radar Measurements of Birds and Micro-Drone," *2016 IEEE Radar Conference*, Philadelphia, PA, 2016, pp. 1–5

[20] Green, J. L., and B. Balsley, "Identification of Flying Birds Using a Doppler Radar," *Proc. Conf. Biol. Aspects Bird/Aircraft Collision Problem*, Clemson University, 1974, pp. 491–508.

[21] Ozcan, A. H., et al., "Micro-Doppler Effect Analysis of Single Bird and Bird Flock for Linear FMCW Radar," *2012 20th Signal Processing and Communications Application Conference*, 2012.

[22] Hoffmann, F., et al., "Micro-Doppler Based Detection and Tracking of UAVs with Multistatic Radar," *Proceedings of 2016 IEEE Radar Conference*, 2016, pp. 1–6.

[23] Kim, B. K., H. -S. Kang, and S. -O. Park, "Experimental Analysis of Small Drone Polarimetry Based on Micro-Doppler Signature," *IEEE Geoscience and Remote Sensing Letters*, Vol. 14, No. 10, 2017, pp. 1670–1674.

[24]   Jian, M., Z. Z. Lu, and V. C. Chen, "Experimental Study on Radar Micro-Doppler Signatures of Unmanned Aerial Vehicles," *Proceedings of 2017 IEEE Radar Conference*, 2017, pp. 854–857.

[25]   Nanzer, J. A., and V. C. Chen, "Microwave Interferometric and Doppler Radar Measurements of a UAV," *Proceedings of 2017 IEEE Radar Conference*, 2017, pp. 1628–1633.

[26]   Ritchie, M., F. Fioranelli, and H. Griffiths, "Micro-Drone RCS Analysis," *Proc. of IEEE Radar Conference*, Johannesburg, South Africa, October 2015, pp. 452–456.

[27]   Goldstein, H., *Classical Mechanics*, 2nd ed., Reading, MA: Addison-Wesley, 1980.

[28]   Trindade, M., and R. Sampaio, "On the Numerical Integration of Rigid Body Nonlinear Dynamics in Presence of Parameters Singularities," *Journal of the Brazilian Society of Mechanical Sciences*, Vol. 23, No. 1, 2001.

[29]   Chen, V. C., C. -T. Lin, and W. P. Pala, "Time-Varying Doppler Analysis of Electromagnetic Backscattering from Rotating Object," *The IEEE Radar Conference Record*, Verona, NY, April 24–27, 2006, pp. 807–812.

[30]   Chen, V. C., "Doppler Signatures of Radar Backscattering from Objects with Micro-Motions," *IET Signal Processing*, Vol. 2, No. 3, 2008, pp. 291–300.

[31]   Persico, A. R., et al., "On Model, Algorithms, and Experiment for Micro-Doppler-Based Recognition of Ballistic Targets," *IEEE Transactions on Aerospace and Electronic Systems*, Vol. 53, No. 3, 2017, pp. 1088–1108.

[32]   Gao, H., et al., "Micro-Doppler Signature Extraction from Ballistic Target with Micro-Motions," *IEEE Transactions on Aerospace and Electronics Systems*, Vol. 46, No. 4, 2010, pp. 1968–1982.

[33]   Lei, P., J. Wang, and J. Sun, "Analysis of Radar Micro-Doppler Signatures from Rigid Targets in Space Based on Inertial Parameters," *IET Radar, Sonar, Navigation*, Vol. 5, No. 2, 2011, pp. 93–102.

[34]   Zhou, Y., "Micro-Doppler Curves Extraction and Parameters Estimation for Cone-Shaped Target with Occlusion Effect," *IEEE Sensors Journal*, 2018.

[35]   Li, M., and Y. S. Jiang, "Feature Extraction of Micro-Motion Frequency and the Maximum Wobble Angle in a Small Range of Missile Warhead Based on Micro-Doppler Effect," *Optics and Spectroscopy*, Vol. 117, No. 5, 2014, pp. 832–838.

[36]   Choi, I. O., "Estimation of the Micro-Motion Parameters of a Missile Warhead Using a Micro-Doppler Profile," *2016 IEEE Radar Conference*, 2016, pp. 1–5.

[37]   Shi, Y. C., et al., "A Coning Micro-Doppler Signals Separation Algorithm Based on Time-Frequency Information," *2017 IEEE International Conference on Signal Processing, Communications and Computing (ICSPCC)*, 2017, pp. 1–5.

[38]   Rashid, L. S., and A. K. Brown, "Impact Modeling of Wind Farms on Marine Navigational Radar," *IET 2007 International Conference on Radar Systems*, Edinburgh, U.K., October 5–18, 2007.

[39]   Casanova, A. C., et al., "Wind Farming Interference Effects," *2008 5th International Multi-Conference on Systems, Signals, and Devices*, July 20–23, 2008.

[40]   Darcy, F., and D. de la Vega, "A Methodology for Calculating the Interference of a Wind Farm on Weather Radar," *2009 Loughborough Antennas & Propagation Conference*, 2009, pp. 665–667.

[41]   Spera, D. A., (ed.), *Wind Turbine Technology*, Ch. 9, New York: The American Society of Mechanical Engineers, 1998.

[42]   *The Effect of Windmill Farms on Military Readiness*, Office of the Director of Defense Research and Engineering, Report to the Congressional Defense Committees, U.S. Department of Defense, 2006.

[43]   Theil, A., and L. J. van Ewijk, "Radar Performance Degradation Due to the Presence of Wind Turbines," *IEEE 2007 Radar Conference*, April 17–20, 2007, pp. 75–80.

[44]   Johnson, K., et al., *Data Collection Plans for Investigating the Effect of Wind Farms on Federal Aviation Administration Air Traffic Control Radar Installations*, Technical Memorandum OU/AEC 05-19TM 00012/4-1, Avionics Engineering Center, Ohio University, Athens, OH, January 2006.

[45]   *Feasibility of Mitigating the Effects of Wind Farms on Primary Radar*, Alenia Marconi Systems Ltd, Report W/14/00623/REP, June 2003.

[46]   Kent, B. M., et al., "Dynamic Radar Cross Section and Radar Doppler Measurements of Commercial General Electric Windmill Power Turbines Part 1: Predicted and Measured Radar Signatures," *IEEE Antennas and Propagation Magazine*, Vol. 50, No. 2, 2008, pp. 211–219.

[47]   Dabis, H. S., "Wind Turbine Electromagnetic Scatter Modeling Using Physical Optics Techniques," *Renewable Energy*, Vol. 16, 1999, pp. 882–887.

[48]   Kong, F., Y. Zhang, and R. Palner, "Radar Micro-Doppler Signature of Wind Turbines," Chapter 12 in *Radar Micro-Doppler Signature: Processing and Applications*, V. C. Chen, D. Tahmoush, and W. J. Miceli, (eds.), Radar Series 34, IET, 2014, pp. 345–381.

[49]   Chatzigeorgiadis, F., and D. Jenn, "A MATLAB Physical-Optics RCS Prediction Code," *IEEE Antennas and Propagation Magazine*, Vol. 46, No. 4, 2004, pp. 137–139.

4

The Micro-Doppler Effect of Nonrigid Body Motion

The nonrigid body is a deformable body; that is, the distance between two points in the body could vary during body motion and thus the shape of the body could be changed. However, as mentioned in Chapter 2, when studying radar scattering from a nonrigid body motion, the body can be modeled as jointly connected rigid links or segments, and a nonrigid body motion can be treated as multiple rigid bodies' motion.

The human gait has been studied in biomedical engineering, sports medicine, physiotherapy, medical diagnosis, and rehabilitation [1]. Motivated by athletic performance analysis, visual surveillance, and biometrics, the methods of how to extract and analyze various human body and limb movements have attracted much attention. The most commonly used method for human body movement analysis uses visual image sequences [2]. However, visual perception of human body motion can be affected by distance, variations in lighting, deformations of clothing, and shadowing. Radar, as an electromagnetic (EM) sensor, has been widely used for detecting targets of interest, measuring their range, and resolving multiple targets in range and velocity. Because of its long-range capability, excellent day and night performance, coherent nature, and ability to penetrate wall and ground, radar has become a useful tool for studying micromotion in humans and animals.

Beside human body motion, animal motion is also an important non-rigid body motion. Compared to human bipedal motion, the four-legged animal's motion has more choices for its feet striking the ground. In 1887, E. Muybridge documented animal locomotion using photography and published a book on animal locomotion that showed how lions, donkeys, dogs, deer, and elephants strode and ran [3]. Later, based on the understanding of animal locomotion, legged machines appeared [4]. Due to the fixed motion pattern, the performance of such legged machines was very limited. Realizing the insufficiency of fixed motion patterns, a better walking machine using controlled legs was constructed [5].

With the understanding of locomotion, the dynamic/kinematic characteristics and movement patterns of human and animal motions became hot topics in computer vision and computer graphics [2]. Point-light displays were used to demonstrate animated patterns of human and animal body and limb movements. Observers can certainly identify human and animal motions through a limited number of animated point-light displays [6].

Radar has proven its ability to detect targets with small radar cross-sections (RCSs), such as humans and animals. However, the methods of how to analyze human or animal dynamic and kinematic characteristics and how to extract motion patterns from radar returns are still challenging.

In most radar range-Doppler imagery, Doppler modulations induced by the target's rotation, vibration, or human body locomotion have been often observed; these show up as characteristic Doppler frequency distributions at those range cells that correspond to the locations of these micromotion sources. Examples of such sources include the rotating antennas on a ship, the rotor blades of a helicopter, the swinging arms and legs of a human, or other oscillatory motion characteristics in a target. To generate a clear radar imagery of a moving target, an effective motion compensation and image autofocusing algorithm must be applied to remove target translational motion and oscillatory motion components for reducing the induced Doppler distributions in the radar imagery.

However, for the purpose of extracting vibration, rotation, or locomotion characteristics in radar returns, their induced Doppler distributions should not be removed. Instead, they should be further exploited. Therefore, the radar micro-Doppler signatures of targets have been studied from experimental observations to theoretical analysis [7–14]. As discussed in Chapter 1, micro-Doppler signatures are represented in a joint time-frequency domain that provides additional information in the time domain to exploit time-varying micro-Doppler characteristics of the rotating or vibrating components in targets. The micro-Doppler characteristics reflect the motion kinematics of a target and provide a unique identification of the target's movement. By carefully analyzing various attributes in the characteristic signature, information about the kinematics of the target can be extracted; this is the basis for discriminating the movement and characterizing the activity of the target. In Chapter 8, the methods of how to analyze micro-Doppler signatures and how to extract component signatures associated with the target's structural component parts will be introduced and discussed.

The radar micro-Doppler signatures of human gait have been investigated since the late 1990s [7, 8]. However, not many works on simulation of quadrupedal animals and their radar micro-Doppler signatures have been published so far. Most of the animal signatures simply illustrate the complicated micro-Doppler signatures of animal motion from collected real radar data. Further studies on the theoretical basis and the simulations of animals' motion are needed.

In this chapter, the biomechanical analysis methods and kinematics of typical nonrigid bodies' motion are introduced. The kinematic model used for describing human motion is given. Based on the kinematic model, micro-Doppler signature analysis for humans can be easily performed. Radar micro-Doppler signatures of simulated and captured human body movements are analyzed in Section 4.1. The micro-Doppler signatures of flapping birds are modeled, simulated, and analyzed in Section 4.2, and quadrupedal animals are introduced in Section 4.3.

4.1   Human Body Articulated Motion

Walking and running are articulated locomotion. The motion of limbs in a human body can be characterized by a repeated periodic movement. The human gait is a highly coordinated periodic movement involving the brain, muscles, nerves, joints, and bones.

Walking is a typical human articulated motion and can be decomposed into a periodic motion in the gait cycle. The human walking cycle consists of two phases: the stance phase and the swing phase. During the stance phase, the foot is on the ground with a heel strike and a toe-off. In the swing phase, the foot is lifted from the ground with acceleration or deceleration. Methods used for human gait analysis can be a visual analysis, sensor measurements, and a kinematic system that measures displacements, velocities, accelerations, orientations of body segments, and angles of joints. Various human body movements, such as walking, running, or jumping, have quite different patterns. As known, radar micro-Doppler signatures are not sensitive to distance, light conditions, and background complexity, which are usually suffered by visual image sequences. The micro-Doppler signature can be easily used to estimate periodicities of gaiting, the period of stance phase, and the period of the swing phase.

4.1.1 Human Walking

The human walk is a periodic motion with each foot from one position of support to the next position of support, periodically swinging arms and legs and moving the body's center of gravity up and down. Even if the human walk has the same general manner, the individual human gait is still carrying personalized characteristics. This is why people can recognize a friend at a distance from his or her walking style [14]. Since deep learning and convolutional neural network have been successfully applied to target classification, micro-Doppler signatures of human gaits may be useful for personal identification through the way a person walks. In addition, the emotional aspects of human gait are often observed in practice. For example, the gait of a cheerful person is quite different from that of a depressed person. Therefore, catching any emotion-like gaiting can help in detecting the anomalous behavior of the person. Another important aspect of human gait analysis is in medical applications, such as medical diagnosis, sports medicine, physiotherapy, and rehabilitation.

Both the dynamic method and the kinematic method can be used to generate human motion. If the motion is created without considering the forces involved, the kinematic method can be easily used to calculate positions of articulated body segments from joint angles as forward kinematics. Then inverse kinematics can be used for determining joint angles from the segments' positions.

Kinematic parameters are the essential parameters of human motion. These parameters include linear position (or displacement), linear velocity, linear acceleration, angular position, angular velocity, and angular acceleration. To completely describe any human motion in a three-dimensional (3-D) Cartesian coordinate system, the linear kinematic parameters of position, velocity, and acceleration define the manner in which the position of any point in the human body changes over time. Velocity is the rate of the position change with respect to time. Acceleration defines the rate of velocity change with respect to time. These three kinematic parameters can be used to understand the motion characteristics of any body movement. If the acceleration can be measured directly with an accelerometer, the corresponding velocity can be estimated by integrating the acceleration, and the corresponding position can be estimated by integrating the velocity.

Angular kinematic parameters include angular position or orientation of body segments (also called segment angles), angular velocity, and angular acceleration. Because the human body is considered by a number of segments linked by joints, the joint angles are very useful parameters. Angular velocity is the rate of angle change with respect to time, and angular acceleration is the rate of angular velocity change with respect to time. These three angular kinematic parameters are used to describe the angular motion of human body segments.

When a rigid body undergoes an angular rotation about an axis, the linear velocity and acceleration of any point in the rigid body can be determined from the angular velocity and acceleration. In the global coordinate system, the angular motion of a rigid body is described by its angular velocity and acceleration. Thus, the linear velocity of a point in the rigid body can be determined by its tangential velocity and normal velocity. The tangential velocity of the point of interest can be derived from the angular velocity and its distance from the rotation center. Then the tangential acceleration of the point is determined by the angular acceleration and the distance from the rotation center. Remember that the tangential and normal velocities and accelerations are given in the body-fixed local coordinate system. By using the rigid body's angular orientation with simple trigonometric identities, the tangential and normal velocities and accelerations can be easily converted to the global coordinate system.

4.1.2 Description of Periodic Motion of Human Walking

The characteristic feature of human walking is its periodicity. Figure 4.1 illustrates the movement of human walking in one cycle [1]. The stance phase occupies about 60% of the cycle, and the swing phase occupies the rest of the

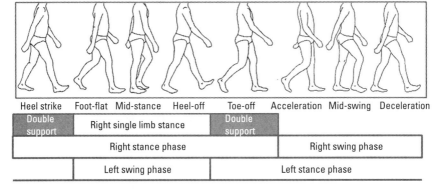

Figure 4.1 Movement of human walking in one cycle.

cycle. In the stance phase, the foot is in contact with the ground. In the swing phase, the foot is lifted from the ground and the leg is swinging and preparing for the next stride. This cyclic movement is repeated over and over again.

The stance phase consists of three periods: (1) first double support, where both feet are in contact with the ground; (2) single limb stance, where only one foot is in contact with the ground and the other foot is swinging; and (3) second double support, where both feet are again in contact with the ground.

During the stance phase, there are five events: heel strike, foot-flat, mid-stance, heel-off, and toe-off. The heel strike initiates the gait cycle, and the toe-off terminates the stance phase because the foot lifts from the ground.

During the swing phase, there is only single-limb swinging. Three events are associated with the swing phase: forward acceleration of the leg, mid-swing when the foot passes directly beneath the body, and deceleration of the leg to stabilize the foot for preparing for the next heel strike.

4.1.3   Simulation of Human Body Movements

To simulate human body movements, a motion model that describes the movement of interest and a human body model that describes the human body parts are needed. The motion model can be a mathematical model or an empirical model. Mathematical modeling consists of constructing a set of equations and simulating the movement of human body segments using a computer. An empirical model is based on large amounts of data on human body motions to formulate empirical human body motion equations and construct a computer model of the human body movement. Then human body model used in biomechanical engineering simplifies the human body and has only as many rigid body segments as needed that are controlled in their movements by joint moments. In the simulation of human body movement, a simplified human body model will be used.

Even if the best way to collect human body movement data is from human subjects, generating human body movement data through computer simulation is still desirable. Simulation allows researchers to study a single parameter isolated from other parameters in the model or to study in conditions where human subjects cannot be tested. Therefore, simulations are important in human body movement studies.

4.1.4   Human Body Segment Parameters

The Denavit-Hartenberg (D-H) convention [15, 16] is a widely used kinematic representation that describes the positions of links and joints in robotics, which can be seen as a simplified case of links and joints in human body parts. The D-H convention states that each link has its own coordinate system with its z-axis in the direction of the joint axis, the x-axis aligned with the outgoing link, and the y-axis orthogonal to the x- and z-axes using a right-handed coordinate system as shown in Figure 4.2.

Once the coordinate systems are determined, the interlink transformations are uniquely described by four parameters: θ is the joint angle about the previous z or z_1 from the old x or x_1 to the new x or x_2; d is the link offset along the previous z to the common normal; a is the length of the common normal; and α is the angle about the common normal, from the old z-axis or z_1 to the new z-axis or z_2 as marked in Figure 4.2. Thus, every link-joint pair is described as a coordinate transform from the previous coordinate system to the next coordinate system.

Any segment in a human body is assumed to be a rigid link, such that the segment size, shape, mass, center of gravity location, and moments of inertia do not change during movements. Under this assumption, a human body is modeled as joints and interconnected rigid links. The movement of such a rigid segment has six degrees of freedom (DOF): three positions in 3-D Cartesian coordinates and three Euler angles of rotation.

The kinematic model of the human body is a hierarchical model of the body links' connectivity, where a family of parent-child spatial relationships are defined and joints become the nodes of the tree structured human body model. In this kinematic tree, all body coordinates are local coordinate systems relative to their parents. Any transformation of a node only affects its

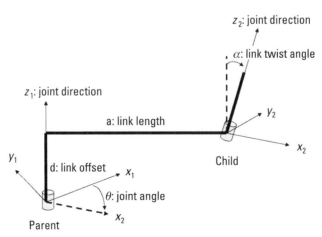

Figure 4.2 The D-H convention used to describe the positions of links and joints.

children nodes, but a transformation of the base (root) node will affect all children nodes in the body tree.

To estimate the kinematic parameters of body segments, Boulic, Magnenat-Thalmann, and Thalmann [17] proposed a global human walking model based on an empirical mathematical parameterization using biomechanical experimental data. This global walking model averages out the personification of walking. This global human walking model will be introduced and used for human gait analysis and study of the micro-Doppler signatures of the human gait.

Motion capture methods use sensors to capture human body movements. Sensors can be active sensors, such as accelerometers, gyroscopes, magnetometers, or passive sensors (e.g., video cameras). The Graphics Laboratory at the Carnegie Mellon University used 12 infrared cameras to capture motions with 41 markers placed on human body segments [18]. The markers' positions and orientations in 3-D space are tracked with a 120-Hz frame rate and stored in a database. In this book, the simulation of radar returns from human walking is based on the empirical mathematical parameterization model [17], and more complex human body movements, such as running, jumping, and other movements, are based on captured motion data provided in the Carnegie Mellon database [18].

4.1.5 Human Walking Model Derived from Empirical Mathematical Parameterizations

Boulic, Magnenat-Thalmann, and Thalmann proposed a global human walk model based on empirical mathematical parameterizations derived from biomechanical experimental data [17]. Because this model is based on averaging parameters from experimental measurements, it is an averaging human walking model without information about personalized motion features. Although the method is for modeling human walking, its principle is also applicable to other human body movements if experimental data is available. In this section, the computer algorithm and the source codes for implementation of this human walking model will be described in detail and will be used to study the micro-Doppler signatures of human walking. To more easily follow Boulic, Magnenat-Thalmann, and Thalmann's paper [17], the symbols used in this section are the same as those used in the paper. The MATLAB source code for the global human walking model is based on [17]. A more detailed description of the empirical equations used in [17] may help to understand the MATLAB simulation of the global human walking model. The source code on the global human walking model is very long because of too many human body segments used. However, it may help readers to understand the whole procedure of the simulation.

This global walking model is derived based on a large number of experimental data but not from solving motion equations. It is intended to provide 3-D spatial positions and orientations of any segment of a walking human body as function of time. Specifically, the motion is described by 12 trajectories, 3 translations, and 14 rotations, five of which are duplicated for both sides of the body, as listed in Table 4.1. These translations and rotations describe one cycle of walking motion (i.e., from right heel strike to right heel strike). They are all dependent on the walking velocity.

According to [17], trajectories are described in three methods. Six trajectories are given by sinusoidal expressions (one of them by a piecewise function), and six trajectories are represented by cubic spline functions passing through control points located at the extremities of these trajectories.

Given a relative walking velocity V_R in m/s (which is normalized by the height of the leg, i.e., rescaled by a dimensionless value of H_l), the relative length of one walking cycle is empirically expressed by $R_C = 1.346 \times V_R$ in meters. Then the time duration of a cycle is defined by $T_C = R_C/V_R$ in seconds, and the relative time is normalized by a dimensionless value of T_C is $t_R = t/T_C$ in

Table 4.1
Body Trajectories

Trajectory	Translation	Body Rotation	Left Rotation	Right Rotation
Vertical translation	$T_V(t)$			
Lateral translation	$T_L(t)$			
Translation forward/backward	$T_{FB}(t)$			
Rotation forward/backward		$\theta_{FB}(t)$		
Rotation left/right		$\theta_{LR}(t)$		
Torsion rotation		$\theta_{TO}(t)$		
Flexing at the hip			$\theta_H(t)$	$\theta_H(t+0.5)$
Flexing at the knee			$\theta_K(t)$	$\theta_K(t+0.5)$
Flexing at the ankle			$\theta_A(t)$	$\theta_A(t+0.5)$
Motion of the thorax		$\theta_{TH}(t)$		
Flexing at the shoulder			$\theta_S(t)$	$\theta_S(t+0.5)$
Flexing at the elbow			$\theta_E(t)$	$\theta_E(t+0.5)$

seconds. The time duration of support is $T_S = 0.752T_C - 0.143$, and the time duration of double support is $T_{DS} = 0.252T_C - 0.143$. The body-fixed local coordinate system is centered at the origin of the spine. The height of the origin of the spine is about 58% of the human height H in meters.

Thus, the translational trajectories are:

1. *Vertical translation:* This is a vertical offset of the center of the spine from the height of the spine. The translation is

$$T_{r_{\text{vertical}}} = -a_v + a_v \sin\left[2\pi\left(2t_R - 0.35\right)\right] \tag{4.1}$$

   where $a_v = 0.015V_R$. The vertical translation function in meter is plotted in Figure 4.3.

2. *Lateral translation:* This is a lateral oscillation of the center of the spine. The translation is

$$T_{r_{\text{lateral}}} = a_1 \sin\left[2\pi\left(t_R - 0.1\right)\right] \tag{4.2}$$

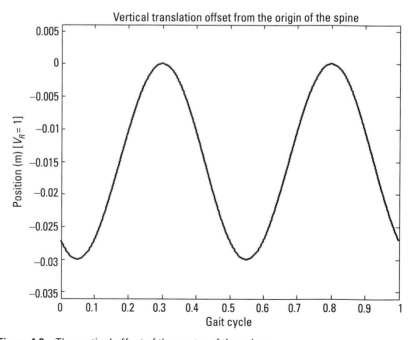

Figure 4.3   The vertical offset of the center of the spine.

where

$$
a_1 = \begin{cases} -0.128V_R^2 + 0.128V_R & \left(V_R < 0.5\right) \\ -0.032 & \left(V_R > 0.5\right) \end{cases}
\tag{4.3}
$$

The lateral translation function is plotted in Figure 4.4.

3. *Translation forward/backward:* This is the body acceleration and deceleration when advancing a new step of a leg and stabilizing the leg. The translation is

$$
T_{r_{\text{F/B}}} = a_{\text{F/B}} \sin\left[2\pi\left(2t_R + 2\varphi_{\text{F/B}}\right)\right]
\tag{4.4}
$$

where

$$
a_{\text{F/B}} = \begin{cases} -0.084V_R^2 + 0.084V_R & \left(V_R < 0.5\right) \\ -0.021 & \left(V_R > 0.5\right) \end{cases}
\tag{4.5}
$$

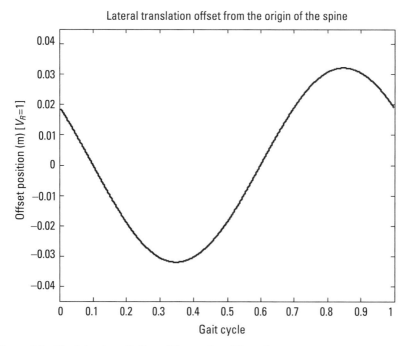

Figure 4.4 The lateral oscillation of the center of the spine.

and $\varphi_{F/B} = 0.625 - T_S$. The translation function is plotted in Figure 4.5.

The three trajectories of rotations are:

1. *Rotation forward/backward:* This is a flexing movement of the back of the body relative to the pelvis before each step to make a forward motion of the leg. The rotation expressed in degree is

$$R_{O_{F/B}} = -ar_{F/B} + ar_{F/B} \sin\left[2\pi\left(2t_R - 0.1\right)\right] \quad (4.6)$$

where

$$ar_{F/B} = \begin{cases} -8V_R^2 + 8V_R & \left(V_R < 0.5\right) \\ 2 & \left(V_R > 0.5\right) \end{cases} \quad (4.7)$$

The rotation function is plotted in Figure 4.6.

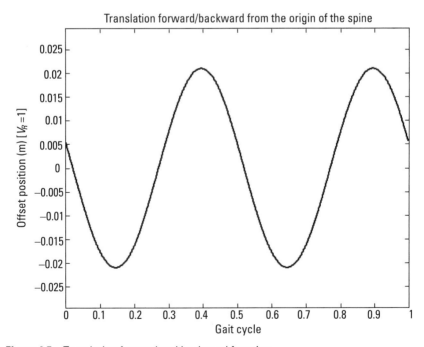

Figure 4.5 Translation forward and backward function.

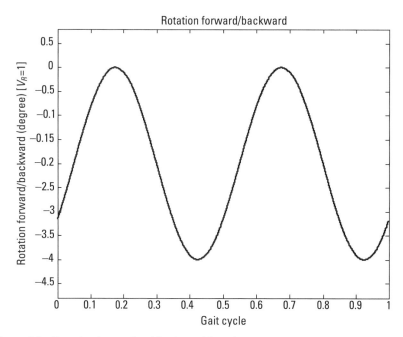

Figure 4.6 Rotation forward and backward function.

2. *Rotation left/right:* This is a flexing movement that makes the pelvis fall on the side of the swinging leg. The piecewise function of the rotation is expressed by

$$R_{O_{L/R}} = \begin{cases} -ar_{L/R} + ar_{L/R}\cos\left[2\pi\left(10t_R/3\right)\right] & \left(0 \leq t_R < 0.15\right) \\ -ar_{L/R} - ar_{L/R}\cos\left\{2\pi\left[\left[10\left(t_R - 0.15\right)/7\right]\right]\right\} & \left(0.15 \leq t_R < 0.5\right) \\ -ar_{L/R} - ar_{L/R}\cos\left\{2\pi\left[\left[10\left(t_R - 0.5\right)/3\right]\right]\right\} & \left(0.5 \leq t_R < 0.65\right) \\ -ar_{L/R} + ar_{L/R}\cos\left\{2\pi\left[\left[10\left(t_R - 0.65\right)/7\right]\right]\right\} & \left(0.65 \leq t_R < 1\right) \end{cases}$$

(4.8)

where $ar_{L/R} = 1.66V_R$. The rotation function is plotted in Figure 4.7.

3. *Torsion rotation:* The pelvis rotates relative to the spine to make a step. The rotation expressed in degree is

$$R_{O_{Tor}} = -ar_{Tor}\cos\left(2\pi t_R\right)$$

(4.9)

where $ar_{Tor} = 4V_R$. The rotation function is plotted in Figure 4.8.

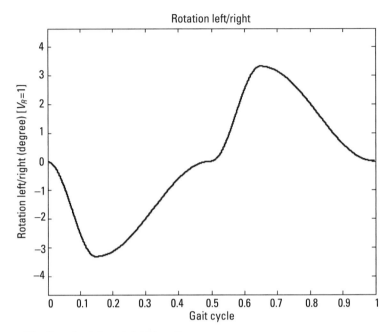

Figure 4.7 Rotation left and right function.

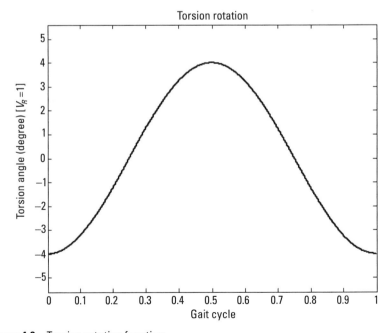

Figure 4.8 Torsion rotation function.

The six trajectories of flexing or torsion in lower body and upper body are:

1. *Flexing at the hip:* There are three control points to be fitted. The flexing function of the hip in degree is plotted in Figure 4.9.
2. *Flexing at the knee:* There are four control points to be fitted. The flexing function of the knee is plotted in Figure 4.10.
3. *Flexing at the ankle:* There are five control points to be fitted. The flexing function of the ankle is plotted in Figure 4.11.
4. *Motion of the thorax:* There are four control points to be fitted. The motion function of the thorax is plotted in Figure 4.12.
5. *Flexing at the shoulder:* The shoulder has left and right rotations and the flexing function is

$$R_{O_{\text{Should}}} = 3 - ar_{\text{Should}} \cos\left(2\pi t_R\right) \qquad (4.10)$$

where $ar_{\text{Should}} = 9.88 V_R$. The flexing function of the shoulder is plotted in Figure 4.13.

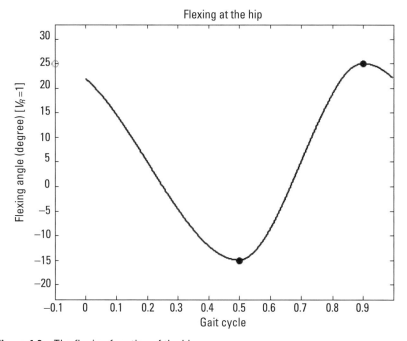

Figure 4.9 The flexing function of the hip.

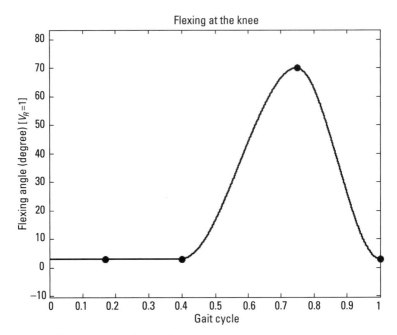

Figure 4.10   The flexing function of the knee.

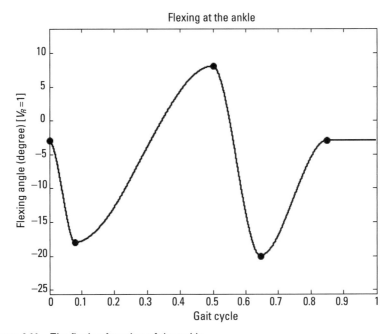

Figure 4.11   The flexing function of the ankle.

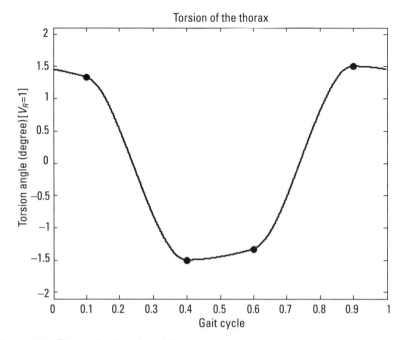

Figure 4.12   The motion function of the torso.

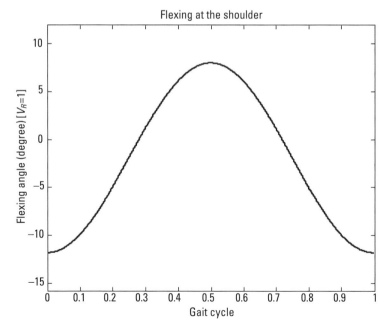

Figure 4.13   The flexing function of the shoulder.

6. *Flexing at the elbow:* The flexing function of the elbow is similar in shape to that of the shoulder, but the flexing angle of the elbow has no negative value. The flexing function is plotted in Figure 4.14.

Because these single equations and piecewise functions are differentiable through a multiple cycle of a time span, the method used to calculate the final result is to form a spline function by placing two additional sets of control points in that there is one cycle before and one cycle after the time span that represents the current cycle. Only the data from the middle cycle is used to guarantee continuity and differentiability when repeated periodically.

Having correctly calculated the necessary movement trajectories, a workable walking model can be developed using these trajectories to calculate the location of a series of 17 reference points on the human body in 3-D space: head, neck, base of spine, left and right shoulders, elbows, hands, hips, knees, ankles, and toes. The (x, y, z)-coordinates are defined as the positive x direction is forward, the positive y direction is right, the positive z direction is up, and the base of the spine is situated at the origin (see Figure 4.15).

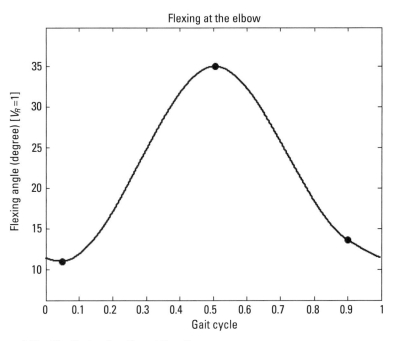

Figure 4.14 The flexing function of the elbow.

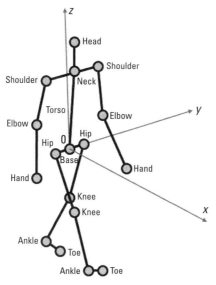

Figure 4.15   Reference points on a human body.

The 3-D orientation of a body segment is determined by locating the 17 joint points based on the body reference coordinates centered at the origin of the spine as illustrated in Figure 4.16. From the flexing angle functions and the translations of the joint points described by the model of biomechanical experimental data, the Euler angle rotation matrix is used to calculate the positions of the 17 joint points at each frame time. By carefully handling the flexing and translation, the 3-D trajectories of these joint points are obtained. These linear and angular kinematic parameters of a walking human are used to simulate radar returns from the human.

In [19] a list of the length of each human body segment normalized by the height of the human was given. The human model used in this book has 17 reference joint points and segment lengths, as shown in Figure 4.17.

To calculate the location of each reference point based on the trajectories, the Euler rotation matrix is applied based on the XYZ convention, where the roll angle is ψ, the pitch angle is θ, and the yaw angle is φ. After the rotation transformations, the location of the reference points can be obtained. To allow for the accurate calculation of body reference points based on multiple angles, the outermost angles must be accounted for first (see Figure 4.18). Translations are then handled after contributions from angles have been accounted for.

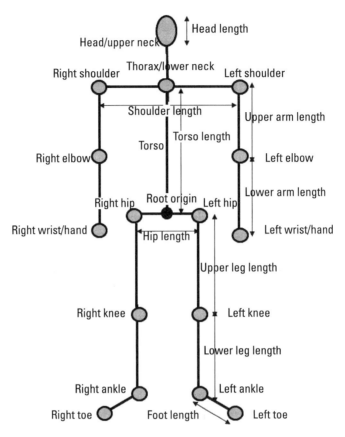

Figure 4.16   The 3-D orientations of human body segments determined by 17 joint points based on the body reference coordinates centered at the origin of the spine (base).

Given a series of reference points over a period of time, animating a model using this data has confirmed the validity of the model. Animating this model shows that the model is able to produce a proper walking human model (see Figure 4.19). This book provides lists of the MATLAB source code to implement the human walking model proposed in [17] and to visualize an animated walking human.

The trajectories of individual body parts of the walking person in part are shown in Figure 4.20 and the corresponding radial velocities are calculated and shown in Figure 4.21. The radial velocity pattern of the person in Figure 4.22 is identical to the micro-Doppler signature of radar backscattering from a walking person shown in Section 4.1.9.

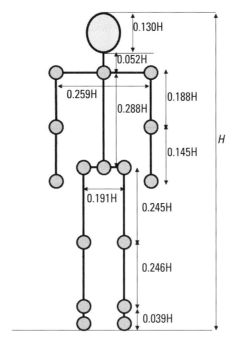

Figure 4.17 The segment lengths used in the human model.

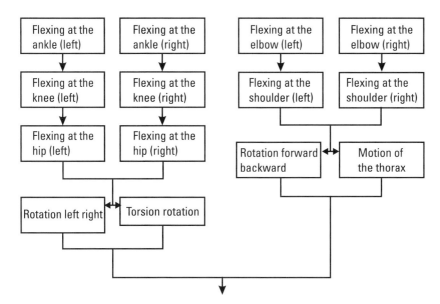

Figure 4.18 The order of the angle trajectory calculation.

Figure 4.19 Animating the human walk.

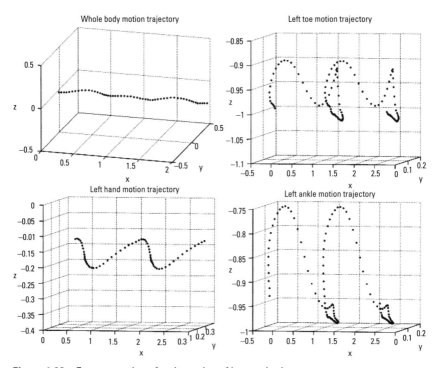

Figure 4.20 Four examples of trajectories of human body parts.

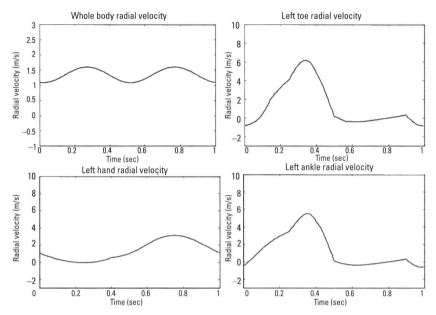

Figure 4.21 Corresponding radial velocities of a human walking toward a radar.

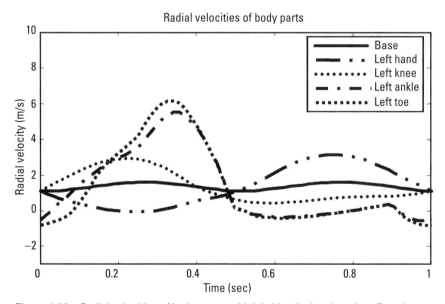

Figure 4.22 Radial velocities of body parts, which is identical to the micro-Doppler signature of a human walking toward a radar.

4.1.6   Capturing Human Motion Kinematic Parameters

To capture human motion, the sensors used can be active or passive. Active sensors transmit signals to a human object and receive the reflected signals from the object. Passive sensors do not transmit any signal and only receive reflected signals from objects illuminated by other sources. Markers used in motion capture systems can be passive or active [20, 21]. Light-point displays are useful active markers that appeared in the 1970s [22] and clearly demonstrated various motion signatures of different animals.

For sensing 3-D motion, commonly used active sensors include accelerometers, gyroscopes, magnetometers, acoustic sensors, and even radar sensors. An accelerometer is a small sensor attached to an object for measuring its acceleration. It measures deflection caused by the movements of the sensor and converts the deflection into an electrical signal. An electromagnetic sensor is attached to joints of any two connected segments and measures the orientation and position of the joint points with respect to the Earth's magnetic field. These active sensors are now widely used to track the positions of joints or segments in a 3-D space.

However, accurate estimates of the corresponding displacement and velocity from the measured acceleration are critical. The velocity is determined by the integral of the acceleration with respect to time, and the displacement is the integral of the velocity with respect to time. For correctly performing the integration process, the measured acceleration histories must be integrated by iteratively adding successive changes in the velocity. Thus, the velocity history and displacement history are computed by

$$
v_i = \frac{a_i + a_{i-1}}{2}\Delta t + v_{i-1}
$$

$$
x_i = \frac{v_i + v_{i-1}}{2}\Delta t + x_{i-1}
$$

(4.11)

where Δt is the time interval between two successive measured acceleration samples, a_i is the measured acceleration at the sampling time i, v_i is the estimated velocity at the sampling time i, x_i is the estimated displacement at the sampling time i, and $i = 1, 2, \ldots, N$, where N is the total number of samples of measured acceleration. The first initial velocity and displacement must be known. For a movement activity starting statically, the first initial velocity and displacement can be set to zeros. Otherwise, they have to be measured by using other methods.

For sensing rotational motion, gyroscopes may be used. By combining a gyroscope with an accelerometer, full 6 DOF data are captured. Human body motion in 3-D space can be completely described by the 6 DOF: linear acceleration along each axis and angular rotation about each axis. Therefore, the combination of the accelerometer and the gyroscope may be used as a complete motion-sensing device for capturing human body movement information.

A commonly used passive sensing for capturing human motion kinematics is optical motion-caption system equipped with multiple cameras to record motions of optical markers attached to the moving parts of the human body to be measured. The orientation and arrangement of multiple cameras are illustrated in Figure 4.23. Each camera captures data in two-dimensional (2-D) coordinates and each optical sensor (marker) can be seen by at least two cameras. From these sets of 2-D coordinates, the motion data in 3-D coordinates can be calculated. Then the direct linear transform [23, 24] may be used to represent markers from 2-D camera coordinates to 3-D space.

4.1.7 Three-Dimensional Kinematic Data Collection

The motion capture (MOCAP) database collected by the Graphics Laboratory at the Carnegie Mellon University is available in the public domain and is very useful for studying human body movements [18]. Twelve infrared cameras with a 120-Hz frame rate are placed around a rectangular area to capture human motion data. Humans wear a jumpsuit with 41 markers on it. The cameras can see the markers in infrared. The images from the 12 cameras are

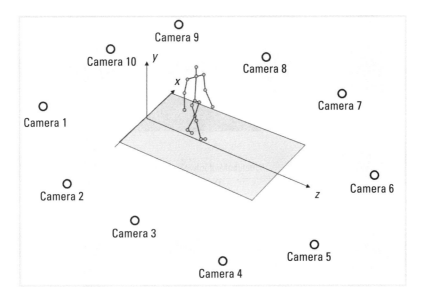

Figure 4.23 Optical motion-caption system equipped with multiple cameras.

processed to generate data of a 3-D human skeleton movement. The skeleton movement data is then stored in a pair of data files. The ASF (skeleton) file in the pair describes information about the skeleton and joints, and the AMC (motion capture) file of the pair contains the movement data.

In the ASF file, the lengths and directions of 30 bone segments are given. There are a total of 30 body segments read out from the AFS file. They are left and right hip joints, left and right femurs, left and right tibias, left and right feet, left and right toes, lower back, upper back, thorax, lower neck, upper neck, head, left and right clavicles, left and right humeri, left and right radii, left and right wrists, left and right hands, left and right fingers, and left and right thumbs. For studying human walking, running, leaping, or jumping, finger and thumb data may not be necessary.

In the motion capture database, the motion of a human object is in the x-z plane with the forwarding direction along the positive z-axis, which is different from the coordinates defined in the human walking model derived in [17] where the human object moves in the x-y plane with the forwarding direction along the positive x direction.

The root of the skeleton hierarchy is a special segment that does not have the direction and length. In the AMC file, the root only contains the starting position and rotation order information. The rotations and directions of other bone segments in the hierarchy are calculated in the AMC file. The rotation of a segment is defined by its axis. The direction of the segment defines the direction from the parent segment to the child segments.

To calculate the global transform for each segment, start by calculating the local transform matrix using the translation offset matrix from its parent segment and the rotation axis matrix. Using the motion data contained in the AMC file and the skeleton defined in the ASF file, the linear and angular kinematic parameters are available for the animation of human body movement and calculating radar returns from human with movements.

Recently, the BVH (BioVision Hierarchy) file format has become a popular and widely used motion data format [25]. The earlier ASF/AMC skeleton and motion capture format used in [18] has been replaced by the BVH format. The entire set of ASF/AMC human motion database collected by the Graphics Laboratory at Carnegie Mellon University has been converted to the BVH format. The BVH file consists of two sections. The first section details the hierarchy and initial pose of the skeleton and the second section describes the degree of freedom data over time or the motion data.

From the captured kinematic database, 3-D motion trajectories of human body segments can be constructed. Figure 4.24 shows a human 2-D position trajectory extracted from the kinematic data of the human root point (i.e.,

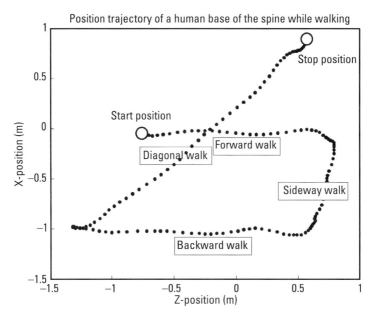

Figure 4.24   The 2-D position trajectory of a walking person's base of the spine when walking forward, then sideway stepping, then walking backward, and finally walking diagonally.

the base of the spine), beginning with walking forward, then sideway stepping, then walking backward, and finally walking diagonally. Figure 4.25 shows the reconstructed animated human model walking sideway stepping and walking backward.

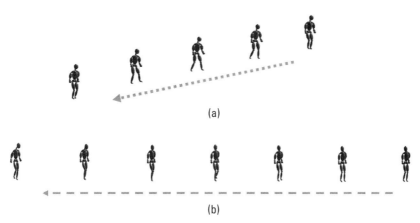

Figure 4.25   Reconstructed animated human model: (a) walking sideway stepping, and (b) walking backward.

However, the human body motion trajectory in Figure 4.24 does not indicate if the human is walking or running, walking forward or walking backward, and walking upstairs or walking downstairs. Besides the reconstructed animated human body movement as shown in Figure 4.25, the use of angle-cyclogram patterns is an additional method to identify detailed human body movement. The angle cyclogram is a phase-space representation of two joint angles.

4.1.8 Characteristics of Angular Kinematics Using the Angle-Cyclogram Pattern

A dynamic system, such as human locomotion, can be described by a set of state variables. The joint angles or joint velocities are these state variables and can be used to represent the human locomotion. Measurable locomotion descriptors, such as the step size, step frequency, and time durations of swing, stance, or double support, are important for the locomotion. To describe any repetitive movement activity, the cyclogram is a useful method [26]. Instead of describing one individual joint kinematics, the cyclogram describes a coordinated movement of two joints linked by two or more segments. The perimeter of the cyclogram is a zero-order moment. The position of the center of mass of the cyclogram is a combination of the zero-order and the first-order moments [26].

The angle-angle cyclogram of human locomotion describes the posture of the leg and the coordination of the hip joint and the knee joint. It is a function of the slope and the speed of human locomotion. However, the angle-angle cyclogram does not describe the velocities involved in the leg (including the femur and the tibia). The angle-velocity phase diagram is a trajectory in the phase space and represents the dynamics of a joint. However, the angle-velocity phase gram has no information about the coordination between two joints. Therefore, the combination of the angle-angle cyclogram with the angle-velocity phase diagram provides informative signatures of human locomotion.

Figure 4.26 depicts the joint angles defined in the human lower body. The hip joint angle can have positive and negative values. However, the knee joint angle can have only a single signed value, for which either a negative sign or a positive sign depends on how the angle is defined.

Figure 4.27 shows an example of joint trajectories (i.e., variations of joint angle with time) of a walking human. The hip and knee joint angles of the walking human are from the motion capture database.

In the cyclogram shown in Figure 4.28, the knee joint angle is plotted against the hip joint angle for a complete walking cycle with arrows that indicate time increases. The backward walking shows the revised time arrows as

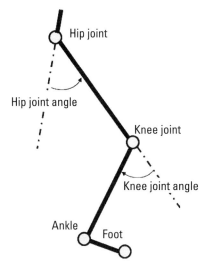

Figure 4.26   The defined joint angles of lower body segments.

shown in Figure 4.29(b). Figure 4.29(c, d) demonstrate different cyclograms of a forward running and a jumping person, respectively.

4.1.9   Radar Backscattering from a Walking Human

Having animated a human body movement model, it is easy to calculate radar back-scattering from the human. POFACET models can be used to calculate the RCSs of human body segments. In this case, human body models should

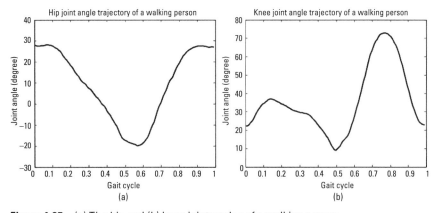

Figure 4.27   (a) The hip and (b) knee joint angles of a walking person.

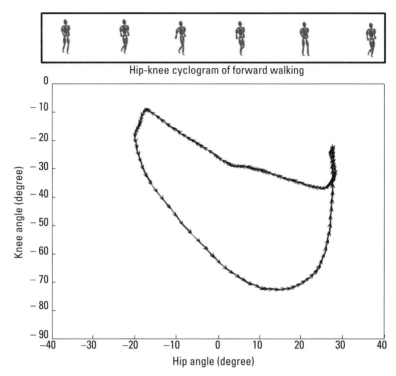

Figure 4.28 The hip-knee cyclogram of a walking person.

be more accurate computer-aided design (CAD) or 3-D graphics models that allow users to build more accurate human bodies.

For simplicity, in this book, human body segments are modeled by ellipsoids. The RCS of an ellipsoid, RCS_{ellip} is given in (3.21), where a, b, and c represent the length of the three semiaxes of the ellipsoid in the x, y, and z directions, respectively. The incident aspect angle θ and the azimuth angle φ represent the orientation of the ellipsoid relative to the radar and are illustrated in Figure 4.30. These RCS formulas are used to simulate radar backscattering from a human with movements. It should be mentioned that if human body segments are modeled by 3-D ellipsoids, the use of POFACET to calculate the RCSs of ellipsoids is not necessary.

Figure 4.31(a) illustrates the geometry of a radar and a walking human, where the radar is located at $(X_1 = 10\text{m}, Y_1 = 0\text{m}, Z_1 = 2\text{m})$ with a wavelength of 0.02m and the starting point of the human base is located at $(X_0 = 0\text{m}, Y_0 = 0\text{m}, Z_0 = 0\text{m})$. The wavelength of the radar is 0.02m. Using the human walking model derived in [17], assume that the relative velocity of the walking

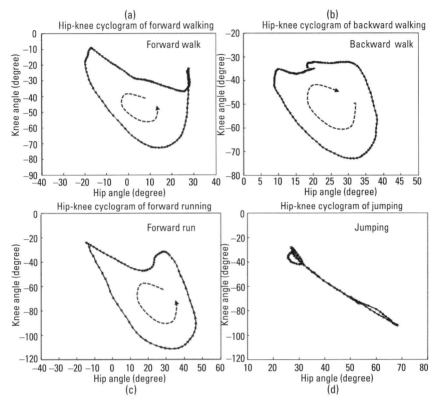

Figure 4.29 The hip-knee cyclogram pattern of: (a) forward walking, (b) backward walking, (c) forward running, and (d) jumping.

person is $V_R = 1.0 \ sec^{-1}$, the height of the person is $H = 1.8m$, and the mean value of the torso velocity of the person is 1.33 m/s. The corresponding Doppler frequency shift at the given wavelength of 0.02m is $2 \times 1.33/0.02 = 133$ Hz.

With the human modeled by ellipsoidal body segments, the radar backscattering from the walking human can be calculated and the 2-D pulse-range profiles are shown in Figure 4.31(b). The micro-Doppler signature derived from the range profiles is shown in Figure 4.31(c), where the micro-Doppler components of the torso, foot, tibia, and clavicle are indicated through the simulation of separated human body segments.

4.1.10 Human Body Movement Data Processing

In measured radar data, due to background objects and unwanted moving objects, range profiles show strong clutter. To extract the useful data in range

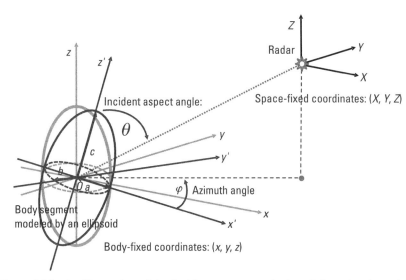

Figure 4.30 An illustration of the incident aspect angle θ and the azimuth angle φ representing the orientation of the ellipsoidal human body segment relative to` the radar.

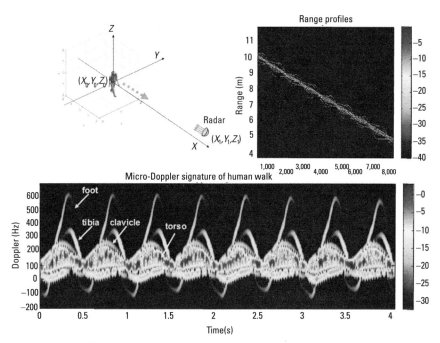

Figure 4.31 (a) The geometry of a radar and a walking human, (b) the radar 2-D pulse-range profiles, and (c) the micro-Doppler signature of the walking human using the model proposed in [17].

profiles, the clutter must be suppressed. Fortunately, most background objects are stationary and the background clutter can be easily suppressed by a notch filter. Radar backscattering from unwanted moving objects may also be filtered as long as they can be distinguished by their ranges and speeds.

4.1.10.1 Clutter Suppression

Clutter suppression techniques utilize statistical properties of radar returns from stationary objects, which are usually near zero mean Doppler frequency and have a smaller bandwidth of the spectrum. The returns from a moving human are offset from the zero Doppler shifts because of its radial velocity. As illustrated in Figure 4.32, a band-reject filter with a notch around zero velocity can reject most of the clutter without affecting the human motion signal as long as the mean velocity of the human motion is larger than the notch width. Figure 4.32(a) illustrates the Doppler spectrum of clutter and the Doppler spectrum of the human motion. Figure 4.32(b) shows the frequency response of the notch filter and Figure 4.32(d) shows the spectrum of the radar range profiles before notch filtering and after notch filtering. Figure 4.32(c) illustrates the Doppler spectrum after clutter suppression and the clutter cleaned range profiles are shown. The efficiency of the clutter suppression method depends on the notch depth and the relative width of the notch as well as the clutter properties. However, the residue from a strong clutter may still produce a significant bias in estimating the Doppler frequency of the motion. The desired average clutter suppression should be more than 40 dB.

4.1.10.2 Time-Frequency Analysis of Clutter Suppressed Data

The clutter suppressed data, as shown in Figure 4.32(c), should be used to calculate the time-frequency micro-Doppler signature of the walking human as shown in Figure 4.33. The micro-Doppler signatures represented in the joint time-frequency domain help to build a more comprehensive signature knowledge database for various micromotion dynamics. The classification and identification of human motion behavior will be based on the micro-Doppler signature knowledge database.

4.1.11 Human Body Movement-Induced Radar Micro-Doppler Signatures

As shown in Figure 4.33, the radar micro-Doppler signature of a walking human is derived by taking the time-frequency transform to the radar range profiles. In the micro-Doppler signature, each forward leg swing appears as large peaks, and the left-leg and right-leg swings complete one gait cycle. The

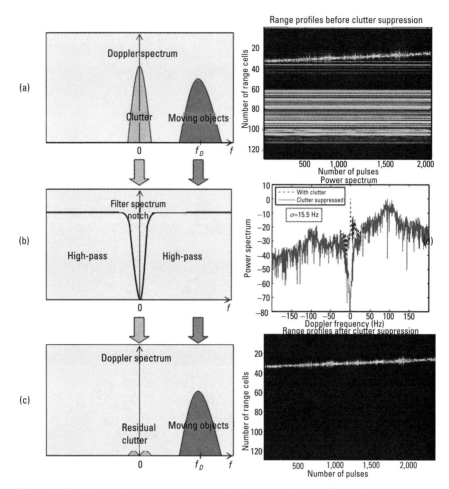

Figure 4.32 Illustration of clutter suppression by a band-reject filter with a notch around zero velocity. (a) Illustration of the Doppler spectrum of clutter and the Doppler spectrum of a human motion. (b) The frequency response of the notch filter. (c) Illustration of the Doppler spectrum after clutter suppression. (d) The spectrum of the radar range profiles before notch filtering and after notch filtering. (*After:* [12].)

body torso motion that is the stronger component underneath the leg swings tends to have a slightly saw-tooth shape because the body speeds up and slows down during the swing as shown.

The micro-Doppler signature is actually an integrated Doppler history of individual body segments during a given observation time. Unlike the captured kinematic data by motion sensing, the radar Doppler history data only carry

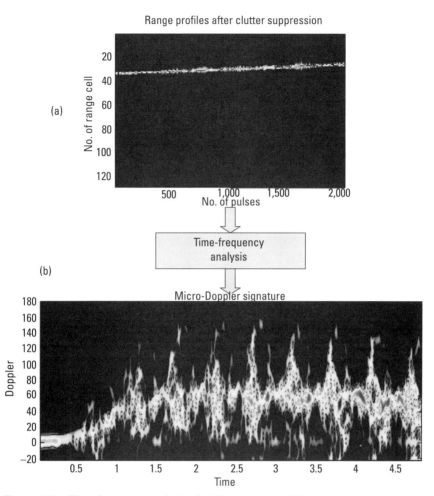

Figure 4.33 Time-frequency analysis of clutter suppressed data.

radial velocity information. Therefore, from the micro-Doppler signatures, the animated body movement model cannot be reconstructed. However, the radar micro-Doppler signatures carry distinctive features of the body movements. It is possible to classify and identify bodies and their movements based on the radar micro-Doppler signatures.

Figure 4.34 shows examples of the radar micro-Doppler signatures of human walking, running, and crawling generated from the collected X-band radar data. Compared to the micro-Doppler signature of a walking person in Figure 4.34(a), the micro-Doppler signature of a running person in Figure

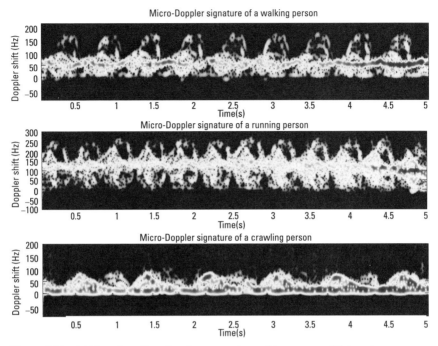

Figure 4.34   (a) The micro-Doppler signature of a walking person, (b) the micro-Doppler signature of a running person, and (c) the micro-Doppler signature of a crawling person.

4.34(b) has a generally higher Doppler frequency shift and a short gait cycle. The crawling person's Doppler frequency shift is much lower and the amplitude of the maximum Doppler shift is also lower, as shown in Figure 4.34(c).

4.1.12   Motion Captured Data for Human Activities

Radar monitoring of human activities in indoor or outdoor environments has been studied for identifying a particular event (such as falling-down or other damaging actions) against other human actions. In general, monitoring of the human daily activity pattern can provide useful information on irregularities and anomalies in the pattern. Micro-Doppler signatures of human activities have been exploited for recognizing human activities for home security/safety, home automation, and health status monitoring [27–33].

Human activities include regular repeated periodic activities (such as walking, running, and swimming) as well as nonperiodic movements (such as standing up, sitting down, kneeling, and falling). Nonperiodic human motion represents another important class of regular human body movement event. Such aperiodic events can be an important indication of health, for example, a chronic limp, concussion, dizziness, or even critical event such as heart attack. Because such micro-Doppler frequencies are directly associated with motion and maneuvering behaviors of the human body parts, these micro-Doppler signatures can be used to characterize and classify such motions and maneuvering patterns. By carefully analyzing various patterns in the signature, features unique to different activities may be identified and served as a basis for discrimination and characterization of human body movement.

The Carnegie Mellon University MOCAP database and other motion capture databases are available for studying radar micro-Doppler signatures of various human activities. In the Carnegie Mellon University MOCAP database, there are up to 30 body parts and a base listed in the BVH files. They are: (1) hip/base, (2) left hip joint, (3) left upper leg, (4) left leg, (5) left foot, (6) left toe base, (7) right hip joint, (8) right upper leg, (9) right leg, (10) right foot, (11) right toe base, (12) lower back, (13) spine, (14) spine 1, (15) neck, (16) neck 1, (17) head, (18) left shoulder, (19) left arm, (20) left forearm, (21) left hand, (22) left finger base, (23) left hand index 1, (24) left thumb, (25) right shoulder, (26) right arm, (27) right forearm, (28) right hand, (29) right finger base, (30) right hand index 1, and (31) right thumb.

The MATLAB code for reading BVH data and display animation is provided in this book. The following examples of human activities demonstrate how the micro-Doppler signature is associated with detailed human body motions.

- *Example 1: Typical spinning-back-kicking in martial arts.* Figure 4.35(a) illustrates a typical martial arts kicking technique: spinning-back-kick. The kick is a back kick with spinning body backwards and then throwing the back kick. When the radar is located at a distance from the back of the human body, spinning backwards while raising the upper leg shows positive Doppler shifts. Then the strong back kick produces a high peak of a negative Doppler shift, as shown in Figure 4.35(b). Finally, recovering from the back kicking has positive Doppler shifts.
- *Example 2: Human falls in indoor environment.* Figure 4.36(a) shows a sequence of a human fall event. Scenarios of human falls in indoor environment are quite various. Falls can happen during walking or running, standing or sitting, or facing downwards or backwards. Identifying a fall and specifying its characteristics are very important in assisted living and elderly care. Currently, fall detection has become an active area of research and development in radar.

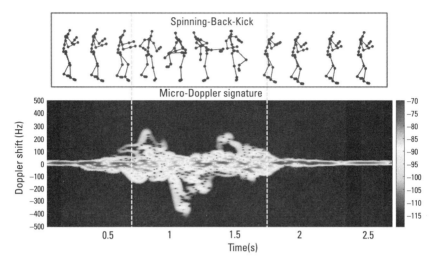

Figure 4.35   (a) A typical kicking technique: spinning-back-kick, and (b) the micro-Doppler signature of spinning, back and kick.

Because fall is an accident, it must have abnormal limb positions and velocities. These abnormal features can be seen in their micro-Doppler signatures. Figure 4.36(b) shows the corresponding micro-Doppler signature of a fall event. The radar is located at a distance from the back of the human body. Thus, the downwards falling shows clearly a high peak with negative Doppler shift. During the falling, some body parts that have abnormal micromotions show positive and negative micro-Doppler shifts.

· *Example 3: Swimming.* Compared to the human walk, the motion of swimming has more choices. The basic types of swimming include backstroke, breaststroke, butterfly stroke, and freestyle. The human walk is a periodic motion with swinging arms and legs. Each cycle of walking consists of two phases: the stance phase and the swing phase. However, each type of swimming may have more phases.

For the backstroke, each arm stroke cycle consists of the entry/extension forward phase, the downsweep phase, the catch phase, the upsweep phase, and the recovery phase. Catch is the phase to apply propulsive force. Upsweep is the propulsive phase of the arm stroke.

Figure 4.37(a) shows a sequence of backstroke. The corresponding micro-Doppler signature in Figure 4.37(b) is the result from micro-motions of 30 body parts. The radar is located at a distance from the back of the swimmer. These positive Doppler peaks are the catch and

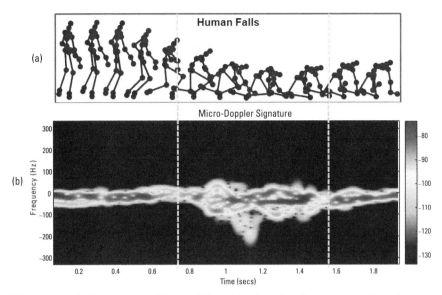

Figure 4.36 (a) A sequence of human fall, and (b) the micro-Doppler signature of the fall event.

upsweep phases. Figure 4.37(c) shows only the isolated micro-Doppler signature of the left forearm for further studying the micromotion feature of each individual body part.

Figures 4.38, 4.39, and 4.40 are micro-Doppler signatures of breaststroke, butterfly stroke, and freestyle swimmers. Different from the backstroke swimming, the arm stroke phases of the breaststroke swimming are outsweep-catch-insweep. The insweep is the propulsive phase of the arm stroke. The butterfly stroke has wave-like body movements and dolphin-kick-like legs' whipping movement. The freestyle swim has downsweep, catch, insweep, and upsweep phases. The propulsive phases are the insweep and upsweep phases.

4.2 Bird Wing Flapping

Bird locomotion is a typical animal locomotion and has attracted much attention for studying avian wings [34, 35]. A bird wing is shown in Figure 4.41, where the wing is defined by the wing span, chord, upper arm (humerus), forearm (ulna and radius), and hand (wrist, hand, and fingers) if using the terms of human arms. Wing flapping is the elevation and depression of the upper

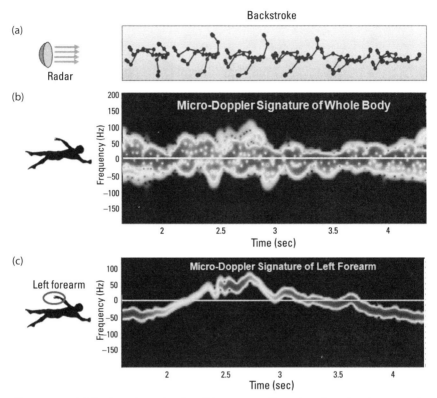

Figure 4.37 (a) Backstroke swimming, (b) corresponding micro-Doppler signature of the backstroke swimming, and (c) the micro-Doppler signature of the left forearm.

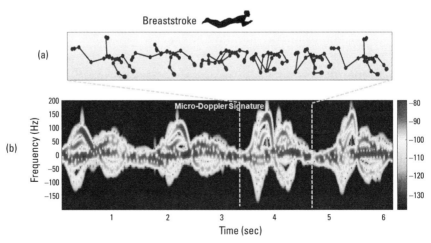

Figure 4.38 (a) Breaststroke swimming, and (b) corresponding micro-Doppler signature of the breaststroke swimming.

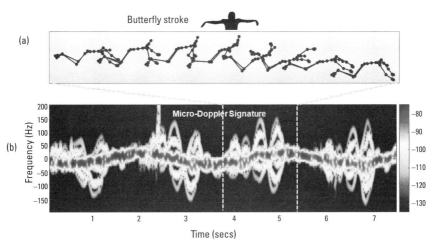

Figure 4.39 (a) Butterfly stroke, and (b) corresponding micro-Doppler signature of the butterfly stroke swimming.

arm or forearm around joints with certain flapping angles. Wing twisting is a rotation of the wing about its principal axis, which results in the elevation of the trailing edge and the depression of the leading edge. Sweeping is the forward protraction or backward retraction of the shoulder. Flapping, twisting, and sweeping consist of the basic bird locomotion.

Wings can have vertical translation, flapping, sweeping, and twisting motions. To study bird wing locomotion, a suitable kinematic model is needed

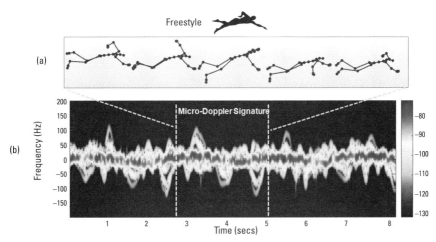

Figure 4.40 (a) Freestyle swimming, and (b) corresponding micro-Doppler signature of the freestyle swimming.

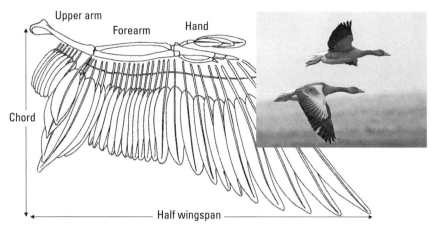

Figure 4.41 Structure of a bird wing.

[36–40]. Having a kinematic model, the bird flight movement can be analyzed. Ramakrishnananda and Wong proposed a model of the forward flapping flight of birds using sophisticated bird geometry with multijointed wings and used defined flapping wing parameters to achieve degrees of freedom [38].

4.2.1 Bird Wing Flapping Kinematics

To calculate bird locomotion, the D-H notation [15] is used to represent joint coordinates for a kinematic chain of revolute and translational joints. When analyzing articulated locomotion, harmonic oscillations are often assumed to describe the style of sinusoidal motion [40]. However, harmonic oscillations are the basis of analyzing complicated motion because any motion can be decomposed into a summation of a series of harmonic components with different amplitudes and frequencies by using the Fourier series.

By defining the flapping angle ψ, the twisting angle θ, and sweeping angle φ, the rotation matrices required for coordinate transformations are

1. Wing flap matrix:

$$\mathfrak{R}_{\text{flap}} = \begin{bmatrix} 1 & 0 & 0 \\ 0 & \cos\psi & \sin\psi \\ 0 & -\sin\psi & \cos\psi \end{bmatrix} \tag{4.12}$$

2. Wing twist matrix:

$$\mathfrak{R}_{\text{twist}} = \begin{bmatrix} \cos\theta & 0 & -\sin\theta \\ 0 & 1 & 0 \\ \sin\theta & 0 & \cos\theta \end{bmatrix} \tag{4.13}$$

3. Wing sweep matrix:

$$\mathfrak{R}_{\text{sweep}} = \begin{bmatrix} \cos\varphi & \sin\varphi & 0 \\ -\sin\varphi & \cos\varphi & 0 \\ 0 & 0 & 1 \end{bmatrix} \tag{4.14}$$

If a wing has a flapping motion, the flapping angle at a given flap frequency f_{flap} is defined as

$$\psi(t) = A_\psi \sin\left(2\pi f_{\text{flap}} t\right) \tag{4.15}$$

the angular velocity is

$$\Omega_\psi(t) = \frac{d}{dt}\psi(t) = 2\pi f_{\text{flap}} A_\psi \cos\left(2\pi f_{\text{flap}} t\right) \tag{4.16}$$

and, thus, the linear velocity of the wing tip is

$$V_\psi(t) = r \cdot \Omega_\psi(t) = 2\pi f_{\text{flap}} \cdot r \cdot A_\psi \cos\left(2\pi f_{\text{flap}} t\right) \tag{4.17}$$

where r is the half wing span.

During the flapping motion, the position of the wing tip in the body-fixed local coordinate system is

$$P_{\text{flap}}(t) = \begin{bmatrix} 0 \\ r \cdot \cos\psi(t) \\ r \cdot \sin\psi(t) \end{bmatrix} \tag{4.18}$$

The linear velocity vector in the body-fixed local coordinate system has only y and z components:

$$V_{\text{flap}}(t) = \begin{bmatrix} 0 \\ V_y(t) \\ V_z(t) \end{bmatrix} = \begin{bmatrix} 0 \\ -2\pi f_{\text{flap}} r A_\psi \cos\left(2\pi f_{\text{flap}} t\right)\sin\psi(t) \\ 2\pi f_{\text{flap}} r A_\psi \cos\left(2\pi f_{\text{flap}} t\right)\cos\psi(t) \end{bmatrix} \tag{4.19}$$

If there is also wing twisting, the rotation transform involves angle flapping first and then angle twisting. With this order of rotations, the position vector of the wing tip is

$$P_{\text{flap-twist}}(t) = \Re_{\text{twist}} \cdot \left(\Re_{\text{flap}} \cdot P_{\text{flap}}(t) \right) =$$

$$\begin{bmatrix} \cos\theta & 0 & -\sin\theta \\ 0 & 1 & 0 \\ \sin\theta & 0 & \cos\theta \end{bmatrix} \cdot \left(\begin{bmatrix} 1 & 0 & 0 \\ 0 & \cos\psi & \sin\psi \\ 0 & -\sin\psi & \cos\psi \end{bmatrix} \cdot P_{\text{flap}}(t) \right)$$

(4.20)

and the velocity vector becomes

$$V_{\text{flap-twist}}(t) = \Re_{\text{twist}} \cdot \left[\Re_{\text{flap}} \cdot V_{\text{flap}}(t) \right] =$$

$$\begin{bmatrix} \cos\theta & 0 & -\sin\theta \\ 0 & 1 & 0 \\ \sin\theta & 0 & \cos\theta \end{bmatrix} \cdot \left(\begin{bmatrix} 1 & 0 & 0 \\ 0 & \cos\psi & \sin\psi \\ 0 & -\sin\psi & \cos\psi \end{bmatrix} \cdot V_{\text{flap}}(t) \right)$$

(4.21)

If the wing has flapping and sweeping, its velocity vector becomes

$$V_{\text{flap-sweep}}(t) = \Re_{\text{sweep}} \cdot \left[\Re_{\text{flap}} \cdot V_{\text{flap}}(t) \right] = \begin{bmatrix} \cos\varphi & \sin\varphi & 0 \\ -\sin\varphi & \cos\varphi & 0 \\ 0 & 0 & 1 \end{bmatrix} \cdot \begin{bmatrix} V_x \\ V_z \sin\psi(t) \\ V_z \cos\psi(t) \end{bmatrix}$$

$$= \begin{bmatrix} V_x \cos\varphi(t) + V_z \sin\psi(t)\sin\varphi(t) \\ -V_x \sin\varphi(t) + V_z \sin\psi(t)\cos\varphi(t) \\ V_z \cos\psi(t) \end{bmatrix}$$

(4.22)

where the flapping angle and the sweeping angle are

$$\psi(t) = A_\psi \sin\left(2\pi f_{\text{flap}} t \right)$$

(4.23)

and

$$\varphi(t) = A_\varphi \cos\left(2\pi f_{\text{sweep}} t \right)$$

(4.24)

where A_ψ and f_{flap} are the amplitude of the flap angle and the flapping frequency, respectively, and A_φ and f_{sweep} are the amplitude of the sweep angle and the sweeping frequency, respectively. The angular velocities are defined by

$$\Omega_\psi(t) = \frac{d}{dt}\psi(t) = 2\pi f_{flap} A_\psi \cos\left(2\pi f_{flap} t\right) \qquad (4.25)$$

and

$$\Omega_\varphi(t) = \frac{d}{dt}\varphi(t) = 2\pi f_{sweep} A_\varphi \sin\left(2\pi f_{sweep} t\right) \qquad (4.26)$$

Thus, the linear velocities at the wing tip are

$$V_\psi(t) = r \cdot \Omega_\psi(t) = 2\pi f_{flap} \cdot r \cdot A_\psi \cos\left(2\pi f_{flap} t\right) \qquad (4.27)$$

and

$$V_\varphi(t) = r \cdot \Omega_\varphi(t) = 2\pi f_{sweep} \cdot r \cdot A_\varphi \sin\left(2\pi f_{sweep} t\right) \qquad (4.28)$$

where r is the half wing span.

A more complicated wing structure may have two segments in the wing. Segment 1 is the upper arm (i.e., the link from the shoulder joint to the elbow joint), and Segment 2 is the forearm (i.e., the link from the elbow to the wrist). For flapping, the elbow joint has only one degree of freedom.

4.2.2 Doppler Observations of Bird Wing Flapping

Radar returns from flying birds carry Doppler modulation caused by flapping wings. The Doppler spread of bird wing flapping has been observed and reported [41–44]. Vaughn [41] cited a report [42] on the investigation of bird classification by using an X-band Doppler radar and showed an 11-second radial velocity history profile of a snowy egret with a 0.97-m wingspan. It showed that the wing flapping rate estimated from the Doppler spectrogram is 4 Hz. From the Doppler spectrogram, the maximum radial velocity expected from an element of the wing can be calculated by

$$\max\{v_{radial}\} = 2A f_{wing} d \qquad (4.29)$$

where A is the amplitude of the wing flapping during the down stroke, f_{wing} is the wing flapping rate, and d is the distance from the bird body center to the tip of the arm. While the bird is taking off, the amplitude A is relatively high (90°–135°). For a snowy egret, the distance d is 0.48m, and the maximum radial velocity should be between 2.1 m/s and 3.7 m/s.

In 1974, Green and Balsley [43] first proposed a method for the identification of flying birds using the time-varying Doppler spectrum (i.e., the micro-Doppler signature). The time-varying Doppler spectrum of a Canada goose shows the power of the radar returned signal from the flying goose at various Doppler frequencies when flying over time. The power spectrum shows that the Doppler shifts from the wings are more than 180 Hz higher than that from the body itself. The bandwidth of the time-varying Doppler spectrum varies with the size of the bird. Thus, the bandwidth may be used to discriminate birds.

Various types of birds, such as passerine-like flapping style or swift-like flapping style, have different patterns of the time-varying Doppler spectrum. The passerine-like flapping style shows repeated clusters of larger fluctuations [44].

4.2.3 Simulation of Bird Wing Flapping

To study the radar returns from a wing-flapping bird, a simple kinematic model with two jointed wing segments is assumed as illustrated in Figure 4.42. In this simulation, the user-defined parameters are the flapping frequency $f_{\text{flap}} = 1.0$ Hz, the length of upper arm $L_1 = 0.5$m, the amplitude of the flapping angle of the upper arm $A_1 = 40°$, the lag of the flapping angle in the upper arm $\psi_{10} = 15°$, the length of forearm $L_2 = 0.5$m, the amplitude of flapping angle in the forearm $A_2 = 30°$, the lag of the flapping angle in the forearm $\psi_{20} = 40°$, and the amplitude of sweeping angle in the forearm $C_2 = 20°$.

With these user-defined parameters, the flapping angle of the upper arm is a harmonic time-varying function given by

$$\psi_1(t) = A_1 \cos\left(2\pi f_{\text{flap}} t\right) + \psi_{10} \tag{4.30}$$

the flapping angle of the forearm is a harmonic time-varying function

$$\psi_2(t) = A_1 \cos\left(2\pi f_{\text{flap}} t\right) + \psi_{20} \tag{4.31}$$

and the twisting angle of the forearm is also a harmonic time-varying function

$$\varphi_2(t) = C_2 \cos\left(2\pi f_{\text{flap}} t\right) + \varphi_{20} \tag{4.32}$$

Therefore, the elbow joint position is $P_1 = [x_1, y_1, z_1]$, where

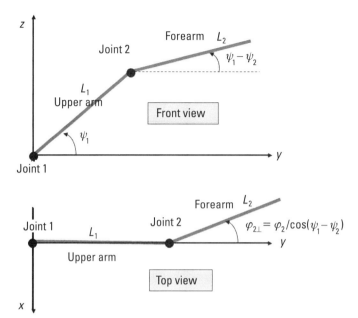

Figure 4.42   A simple kinematic model of a bird wing with two jointed wing segments.

$$x_1(t) = 0;$$

$$y_1(t) = L_1 \cos\left[\frac{\psi_1(t) \cdot \pi}{180}\right]; \qquad (4.33)$$

$$z_1(t) = y_1(t) \cdot \tan\left[\frac{\psi_1(t) \cdot \pi}{180}\right]$$

and the wrist joint position is $P_2 = [x_2, y_2, z_2]$, where

$$x_2(t) = -\left[y_2(t) - y_1(t)\right] \cdot \tan(d);$$

$$y_2(t) = L_1 \cos\left[\frac{\psi_1(t) \cdot \pi}{180}\right] + L_2 \cos\varphi_2(t) \cdot \cos\left[\psi_1(t) - \psi_2(t)\right]; \quad (4.34)$$

$$z_2(t) = z_1(t) + \left[y_2(t) - y_1(t)\right] \cdot \tan\left\{\frac{[\psi_1(t) - \psi_2(t)] \cdot \pi}{180}\right\},$$

where $d = \varphi_2(t)/\cos[\psi_1(t) - \psi_2(t)]$.

With this kinematic model of bird wing locomotion, the simulated bird flying with wing flapping and sweeping is reconstructed. Figure 4.43 is the simulation results, where Figure 4.43(a) shows the wing flapping and sweeping, Figure 4.43(b) is the flight trajectory of two wing tips, and Figure 4.43(c) is the animated flying bird model. Having the simulation, the radar backscattering from the simulated flapping bird can be calculated. Assume that the X-band radar is located at $X = 20$m, $Y = 0$m, and $Z = -10$m, and the bird is flying with a velocity 1.0 m/s. Figure 4.44 is the geometry of the radar and the flying bird. The radar range profiles are shown in Figure 4.45(a) and the micro-Doppler signature of the flying bird with flapping wings is shown in Figure 4.45(b). From the micro-Doppler signature, both the flapping upper arms and forearms with the flapping frequency 1.0 Hz can be seen. The MAT-LAB source codes for the flying bird simulation are provided in this book.

4.3 Quadrupedal Animal Motion

Quadrupedal animals use four-legged motion and thus have more choices on their feet striking the ground than humans. They could strike ground with

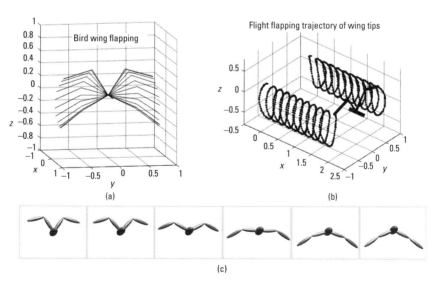

Figure 4.43 A simulation flying bird with a simple kinematic model. (a) The wing flapping and sweeping, (b) the flight trajectory of two wing tips, and (c) the animated flying bird model.

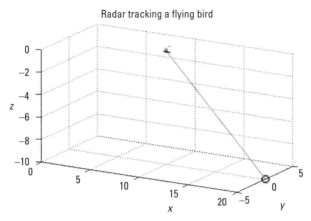

Figure 4.44 The geometry of a radar and flying bird.

each foot separately (four-beat gaits), with two feet separately and other two feet together (three-beat gaits), with three feet together and the other one separately (two-beat gaits), they could strike in pairs (two-beat gaits), or they could strike with all four feet together (one-beat gait). The normal four-legged animals' walking sequence is four evenly spaced beats with left hind, left fore, right hind, and right fore and without a suspension phase.

Muybridge found that all mammals followed the footfall sequence of the horse when walking on four legs. Figure 4.46 shows walking horses. For a half-sequence of horse walking, beginning with an end of a three-legged support phase and the right front leg taking off the ground with the left rear leg pushing back, walking becomes alternately a two-legged support phase, a three-legged support phase, and a two-legged support phase. The other half of the walking sequence is exactly the same as the first half, but with the right and left legs reversed [3].

To study fast animal movements, a high-speed video recorder should be used for recording 3-D motion information, just like the recording of human body movements. At least two synchronized video recorders set at an angle to each other are needed for capturing 3-D motion information. More video recorders arranged separately around the animal movement area can avoid occlusions between animal body parts.

The point-light displays for human body movement study were also used in studying animal body movements [3]. It has been demonstrated that naive observers can identify animals through animated point-light displays [6].

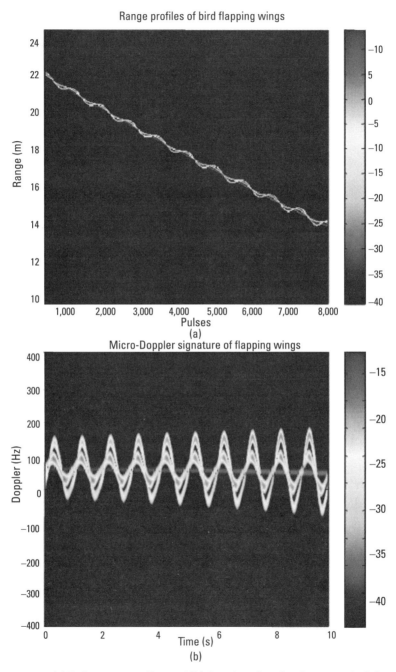

Figure 4.45   (a) Radar range profiles, and (b) the micro-Doppler signature of a flying bird with flapping wings.

Figure 4.46 Walking horses.

4.3.1 Modeling of Quadrupedal Locomotion

As described the hierarchical human kinematic model in Section 4.1, a hierarchical quadrupedal model of animal body links' connectivity can also be used, where a family of parent-child spatial relationships between joints is defined. Figure 4.47 is a quadrupedal model of a dog, where 25 joints are selected. However, a mathematical parameterization modeling of a dog or other four-legged animal body segments is not available. Some commercial motion-captured databases for quadrupedal animals are available. When the

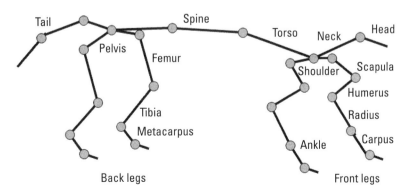

Figure 4.47 A hierarchical quadrupedal model of a dog.

kinematic parameters of a quadrupedal animal are available, with the hierarchical quadrupedal model of animal body links, it is easy to reconstruct an animated quadrupedal animal movement.

From the flexing angle functions and the translations of the joint points described by the model, the positions of the 25 joint points can be calculated at each frame time. These linear and angular kinematic parameters of a quadrupedal animal are used to simulate radar returns from the animal locomotion.

4.3.2 Micro-Doppler Signatures of Quadrupedal Locomotion

Quadrupedal animal motion can be easily distinguished from bipedal human motion by visualization. Radar micro-Doppler signatures of quadrupedal animals are quite different from the micro-Doppler signatures of bipedal humans because four legged animal locomotion can have four-beat gaits, three-beat gaits, two-beat gaits, or one-beat gait. Therefore, micro-Doppler components from legs' motion are much more complicated than the micro-Doppler components of human bipedal motion. Figure 4.48 shows the micro-Doppler signature of a horse carrying a rider walking away from the radar [45]. From

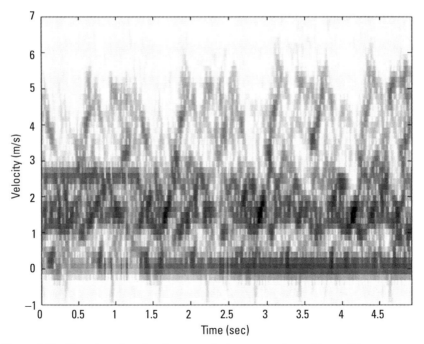

Figure 4.48   The micro-Doppler signature of a horse carrying a rider walking away from a radar. (*After:* [45].)

the micro-Doppler signature, the estimated horse walking radial speed is about 1.5 m/s, and one walking gait cycle is about 1.3 seconds.

Figure 4.49 shows the micro-Doppler signature of a trotting horse carrying a rider [45]. From the micro-Doppler signature, the estimated horse radial speed is about 3 m/s, and one trotting cycle is about 0.8 second. The micro-Doppler signature of a dog approaching the X-band multiple frequency continuous wave radar was also reported in [46].

4.3.3 Summary

Compared to human bipedal motion, the four-legged animals' quadrupedal motion has more choices on feet striking. Thus, motion patterns of animal are more complicated. Unlike the global human walking model derived from a large number of experimental data, there is no similar model available for modeling a four-legged animal walking. Most of the radar micro-Doppler signatures of horses and dogs were generated from collected radar data. To further investigate radar scattering from animals with quadrupedal motion, a model of the quadrupedal locomotion is needed. This model can be either derived

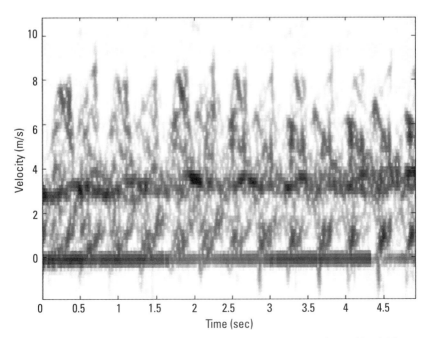

Figure 4.49   The micro-Doppler signature of a trotting horse carrying a rider. (*After:* [45].)

from experimental data or formulated using collected data by motion capture sensors. High-speed cinematographic technique is one of the best methods of recording an animal's gait patterns. To completely describe quadrupedal locomotion in a 3-D Cartesian coordinate system, kinematic parameters are needed to provide the position of any point in the animal body changes over time. These kinematic parameters can be used to understand locomotion characteristics of the quadrupedal movement.

After defining a suitable motion model for modeling quadrupedal locomotion, extracting the motion features of body component parts from the micro-Doppler signatures and identifying quadrupedal gaiting become feasible.

References

[1]   Vaughan, C. L., B. L. Davis, and J. C. O'Connor, *Dynamics of Human Gait*, 2nd ed., Cape Town, South Africa: Kiboho Publishers, 1999.

[2]   Nixon, M. S., and J. N. Carter, "Automatic Recognition by Gait," *Proc. IEEE*, Vol. 94, No. 11, 2006, pp. 2013–2024.

[3]   Muybridge, E., *Animal Locomotion*, Mineola, NY: Dover Publications, 1957 (original work published 1887).

[4]   Raibert, M. H., "Legged Robots," *Communications of the ACM*, Vol. 29, No. 6, 1986, pp. 499–514.

[5]   Liston, R. A., and R. S. Mosher, "A Versatile Walking Truck," *Proceedings of the Transportation Engineering Conference*, Institution of Civil Engineering, London, 1968, pp. 255–268.

[6]   Mather, G., and S. West, "Recognition of Animal Locomotion from Dynamic Point-Light Displays," *Perception*, Vol. 22, No. 7, 1993, pp. 759–766.

[7]   Chen, V. C., "Analysis of Radar Micro-Doppler Signature with Time-Frequency Transform," *Proc. of the IEEE Workshop on Statistical Signal and Array Processing (SSAP)*, Pocono, PA, 2000, pp. 463–466.

[8]   Baker, C. J., and B. D. Trimmer, "Short-Range Surveillance Radar Systems," *Electronics & Communication Engineering Journal*, August 2000, pp. 181–191.

[9]   Geisheimer, J. L., W. S. Marshall, and E. Greneker, "A Continuous-Wave (CW) Radar for Gait Analysis," *35th IEEE Asilomar Conference on Signal, Systems and Computers*, Vol. 1, 2001, pp. 834–838.

[10]   van Dorp, P., and F. C. A. Groen, "Human Walking Estimation with Radar," *IEE Proceedings—Radar, Sonar, and Navigation*, Vol. 150, No. 5, 2003, pp. 356–365.

[11]   Chen, V. C., et al., "Micro-Doppler Effect in Radar: Phenomenon, Model, and Simulation Study," *IEEE Transactions on Aerospace and Electronic Systems*, Vol. 42, No. 1, 2006, pp. 2–21.

[12] Chen, V. C., "Doppler Signatures of Radar Backscattering from Objects with Micro-Motions," *IET Signal Processing*, Vol. 2, No. 3, 2008, pp. 291–300.

[13] Chen, V. C., "Detection and Tracking of Human Motion by Radar," *IEEE 2008 Radar Conference*, Rome, Italy, May 26–29, 2008, pp. 1957–1960.

[14] Cutting, J., and L. Kozlowski, "Recognizing Friends by Their Walk: Gait Perception Without Familiarity Cues," *Bulletin of the Psychonomic Society*, Vol. 9, 1977, pp. 353–356.

[15] Denavit, J., and R. S. Hartenberg, "A Kinematic Notation for Lower-Pair Mechanisms Based on Matrices," *Trans. ASME J. Appl. Mech.*, Vol. 23, 1955, pp. 215–221.

[16] Hartenberg, R. S., and J. Denavit, *Kinematic Synthesis of Linkages*, New York: McGraw-Hill, 1964.

[17] Boulic, R., N. Magnenat-Thalmann, and D. Thalmann, "A Global Human Walking Model with Real-Time Kinematic Personification," *The Visual Computer*, Vol. 6, No. 6, 1990, pp. 344–358.

[18] Motion Research Laboratory, Carnegie Mellon University, http://mocap.cs.cmu.edu.

[19] Winter, D. A., *The Biomechanics and Motor Control of Human Movement*, 2nd ed., New York: John Wiley & Sons, 1990.

[20] Allard, P., I. A. F. Stokes, and J. P. Blanchi, "Three Dimensional Analysis of Human Movement," *Human Kinetics*, 1995.

[21] Bregler, C. and J. Malik, "Tracking People with Twists and Exponential Maps," *International Conference on Computer Vision and Pattern Recognition*, Santa Barbara, CA, 1998.

[22] Johnsson, G., "Visual Motion Perception," *Scientific American*, June 1975, pp. 76–88.

[23] Abdel-Aziz, Y. I., and H. M. Karara, "Direct Linear Transformation from Comparator Coordinates into Object Space Coordinates in Close-Range Photogrammetry," *Proceedings of the Symposium on Close-Range Photogrammetry*, Falls Church, VA: American Society of Photogrammetry, 1971, pp. 1–8.

[24] Miller, N. R., R. Shapiro, and T. M. McLaughlin, "A Technique for Obtaining Spatial Kinematic Parameters of Segments of Biomechanical Systems from Cinematographic Data," *J. Biomech.*, Vol. 13, 1980, pp. 535–547.

[25] Meredithm N. and S. Maddock, "Motion Capture File Formats Explained," www.dcs.shef.ac.uk/intranet/research/resmes/CS0111.pdf.

[26] Debernard, S., et al., "A New Gait Parameterization Technique by Means of Cyclogram Moments: Application to Human Slope Walking," *Gait and Posture*, Vol. 8, No. 1, August 1998, pp. 15–36.

[27] Ram, S. S., S. Z. Gurbuz, and V. C. Chen, "Modeling and Simulation of Human Motion for Micro-Doppler Signatures," Chapter 3 in *Radar for Indoor Monitoring: Detection, Classification, and Assessment*, M. G. Amin, (ed.), Boca Raton, FL: CRC Press/Taylor & Francis Group, 2018, pp. 39–69.

[28] Zhang, Y. M., and D. K. C. Ho, "Continuous-Wave Doppler Radar for Fall Detection," Chapter 4 in *Radar for Indoor Monitoring: Detection, Classification, and Assessment*, M. G. Amin, (ed.), Boca Raton, FL: CRC Press/Taylor & Francis Group, 2018, pp. 71–93.

[29] Wu, Q., et al., "Radar-Based Fall Detection Based on Doppler Time-Frequency Signatures for Assisted Living," *IET Radar, Sonar and Navigation*, Vol. 9, No. 2, 2015, pp. 164–172.

[30] Li, C., et al., "A Review on Recent Advances in Doppler Radar Sensors for Noncontact Healthcare Monitoring," *IEEE Transactions on Microwave Theory and Techniques*, Vol. 61, No. 5, 2013, pp. 2046–2060.

[31] Fioranelli, F., M. Ritchie, and H. Griffiths, "Bistatic Human Micro-Doppler Signatures for Classification of Indoor Activities," *Proceedings of IEEE 2017 Radar Conference*, 2017, pp. 610–615.

[32] Gurbuz, S. Z., et al., "Micro-Doppler-Based In-Home Aided and Unaided Walking Recognition with Multiple Radar and Sonar Systems," *IET Radar, Sonar and Navigation*, Vol. 11, No. 1, 2017, pp. 107–115.

[33] Chen, Q. C., et al., "Joint Fall and Aspect Angle Recognition Using Fine-Grained Micro-Doppler Classification," *Proceedings of IEEE 2017 Radar Conference*, 2017, pp. 912–916.

[34] Colozza, A., "Fly Like a Bird," *IEEE Spectrum*, Vol. 44, No. 5, 2007, pp. 38–43.

[35] Liu, T., et al., "Avian Wings," *The 24th AIAA Aerodynamic Measurement Technology and Ground Testing Conference*, Portland, OR, AIAA Paper No. 2004-2186, June 28–July 1, 2004.

[36] Liu, T., et al., "Avian Wing Geometry and Kinematics," *AIAA Journal*, Vol. 44, No. 5, May 2006, pp. 954–963.

[37] Tobalske, B. W., T. L. Hedrick, and A. A. Biewener, "Wing Kinematics of Avian Flight Across Speeds," *J. Avian Biol.*, Vol. 34, 2003, pp. 177–184.

[38] Ramakrishnananda, B., and K. C. Wong, "Animated Bird Flight Using Aerodynamics," *The Visual Computer*, Vol. 15, 1999, pp. 494–508.

[39] DeLaurier, J. D., and J. M. Harris, "A Study of Mechanical Flapping-Wing Flight," *Aeronautical Journal*, Vol. 97, October 1993, pp. 277–286.

[40] Parslew, B., "Low Order Modeling of Flapping Wing Aerodynamics for Real-Time Model Based Animation of Flapping Flight," Dissertation, School of Mathematics, University of Manchester, 2005.

[41] Vaughn, C. R., "Birds and Insects as Radar Targets: A Review," *Proc. of the IEEE*, Vol. 73, No. 2, 1965, pp. 205–227.

[42] Martison, L. W., A *Preliminary Investigation of Bird Classification by Doppler Radar*, RCA Government and Commercial Systems, Missile and Surface Radar Division, Moorestown, NJ, prepared for NASA Wallops Station, Wallops Island, VA, February 20, 1973.

[43] Green, J. L., and B. Balsley, "Identification of Flying Birds Using a Doppler Radar," *Proc. Conf. Biol. Aspects Bird/Aircraft Collision Problem*, Clemson University, 1974, pp. 491–508.

[44] Zaugg, S., et al., "Automatic Identification of Bird Targets with Radar Via Patterns Produced by Wing Flapping," *Journal of the Royal Society Interface*, 2008.

[45] Tahmoush, D., J. Silvious, and J. Clark, "An UGS Radar with Micro-Doppler Capabilities for Wide Area Persistent Surveillance," *Proceedings of the SPIE, Radar Sensor Technology XIV*, Vol. 7669, 2010, pp. 766904–766911.

[46] Anderson, M. G., and R. L. Rogers, "Micro-Doppler Analysis of Multiple Frequency Continuous Wave Radar Signatures," *Proc. of SPIE, Radar Sensor Technology XI*, Vol. 6547, 2007.

5

Application to Vital Sign Detection

Human vital signs are the most important signs that indicate the status of human body's functions. Although there are four standard vital signs in a medical context (body temperature, heart rate, respiratory rate, and blood pressure), only respiratory rate and heart rate are of concern in emergency situations.

Noncontact vital sign detection is especially needed in health-care monitoring, medical surveillance, and finding survivors in emergency situations. One of the emerging applications of radar sensors is remotely detection of vital signs for monitoring patients in hospitals, finding survivors after earthquakes, and in other emergency situations [1–3].

J. C. Lin first proposed a microwave technique for measuring respiratory movements of human and animal [4]. K. M. Chen et al. developed an X-band life-detection radar system for detecting heartbeat and breathing of human subjects [5, 6]. The detection of vital signs by radars is based on a fact that radar signals returned from a human body are modulated by the tiny physical movements due to breathing and heart beating. Therefore, how to extract useful vital signs from radar signals has become an emerging research topic, including the detection of human heartbeats, thorax motions due to breathing, and even vibration of the larynx.

The information to be captured by Doppler radars may include the respiratory rate, the pulse rate, heartbeat patterns, and respiratory patterns. The respiration rate and its patterns reveal the respiratory physiology. An irregular pulse rate indicates cardiac abnormality. Monitoring the changes in the heart rate and respiratory rate also helps to diagnose sleep apnea.

5.1 Vibrating Surface Modeling of Vital Signs

The detection of vital signs by radar is based on the fact that signals returned from a vibrating surface are modulated by the physical vibration. The human chest wall is such a vibrating surface caused by breathing and heart beating. Figure 5.1 shows a model of radar signals returned from a vibrating surface.

If the vibration frequency of the surface is f_v and the maximum amplitude of the vibration is D_{max}, the vibration function can be simply written by

$$D_v(t) = D_{max} \sin\left(2\pi f_v t\right) \tag{5.1}$$

Assuming the distance from the surface to the radar is R_0 and the radar incident angle to the surface is zero, then the distance between the radar and the vibrating surface becomes

$$R(t) = R_0 + D_{max} \sin\left(2\pi f_v t\right) \tag{5.2}$$

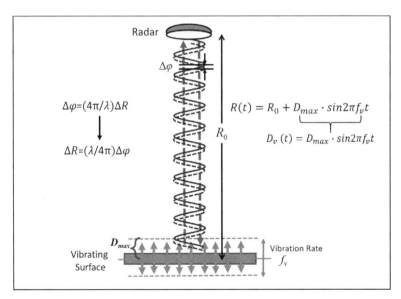

Figure 5.1 Modeling of radar signals returned from a vibrating surface.

For a continuous-wave (CW) radar, the returned signal from the vibrating surface is

$$s_R(t) = A_R \cdot \exp\left\{ j\left[2\pi f_0 t + 4\pi \frac{R(t)}{\lambda} \right] \right\}$$ (5.3)

where A_R is the amplitude, f_0 is the carrier frequency of the transmitted signal, and λ is the wavelength. The phase term $4\pi R(t)/\lambda$ can be rewritten as

$$\varphi(t) = \frac{4\pi}{\lambda} R(t) = \frac{4\pi}{\lambda} R_0 + \frac{4\pi}{\lambda} D_{max} \sin\left(2\pi f_v t\right)$$ (5.4)

and the received signal becomes

$$s_R(t) = A_R \exp\left\{ j\frac{4\pi}{\lambda} R_0 \right\} \exp\left\{ j2\pi f_0 t + \frac{4\pi}{\lambda} D_{max} \sin 2\pi f_v t \right\}$$ (5.5)

By defining a phase modulation function:

$$\theta(t) = \frac{4\pi}{\lambda} D_v(t) = \frac{4\pi}{\lambda} D_{max} \sin 2\pi f_v t$$ (5.6)

the received signal can be expressed by the phase modulation function:

$$s_R(t) = A_R \exp\left\{ j\frac{4\pi}{\lambda} R_0 \right\} \exp\left\{ j2\pi f_0 t + \theta(t) \right\}$$ (5.7)

The vibration induced micro-Doppler shift is the time derivative of the phase modulation function

$$f_{mD}(t) = \frac{1}{2\pi} \frac{d}{dt} \theta(t) = \frac{4\pi}{\lambda} f_v D_{max} \cos\left(2\pi f_v t\right)$$ (5.8)

and the line-of-sight velocity becomes

$$v = \frac{\lambda}{2} f_{mD}(t) = 2\pi f_v D_{max} \cos\left(2\pi f_v t\right)$$ (5.9)

The sensitivity of a radar for noncontact sensing of a surface vibration is as good as other high-sensitive contact sensors such as geophones, which are very sensitive devices used in seismology to record the waves reflected by the vertical motion of the Earth's surface. An experimental measurement was conducted using a K-band CW Doppler radar operating at 25 GHz and located at $R_0 = 0.86$m from a surface vibrating at a frequency of 5.2 Hz. By setting a maximum amplitude of the vibration as low as $D_{max} = 0.005$ mm = 5 micron, the vibration function can still be observed from the collected radar data. The measured power spectrum of the vibration function is shown in Figure 5.2. It shows that the detected vibration signal is about 10 dB more than other peaks due to noise. This measurement indicates that Doppler radars have great potential for noncontact and remote sensing of very weak vibrations for a variety of applications.

For vital signs, the vibration frequency f_v is usually very low. The normal respiration rate of an adult person at rest is about 12 to 16 breaths per minute, which leads to the vibration frequency of the chest wall $f_v \approx 0.2$ to 0.3Hz. A normal heart rate ranges from 60 to 100 beats per minute and, thus, the vibration frequency of the chest wall due to heart beat is $f_v \approx 1$ to 1.7 Hz. The maximum displacement, D_{max}, of the surface vibrating due to the respiration is about 4 to 12 mm and that due to heartbeat is about 0.25 to 0.5 mm, which are large enough compared with the maximum amplitude of the vibration $D_{max} = 0.005$ mm shown in Figure 5.2.

5.2   Homodyne Doppler Radar Systems for Vital Sign Detection

Any Doppler radar system, including the narrowband CW radar and the wideband frequency modulated continuous wave (FMCW) radar, can be used

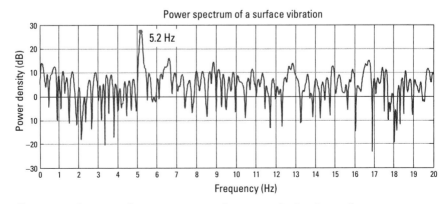

Figure 5.2   A measured power spectrum of a very weak vibrating surface.

for the vital sign detection. The radar receiver can be a homodyne receiver or a heterodyne receiver [7]. An overview of radar system architectures can be found in Chapter 7. Homodyne receiver is a zero-intermediate-frequency (IF) receiver that directly converts radio frequency (RF) signals to baseband signals. However, a heterodyne receiver is not a direct zero-IF one; it has nonzero IF and uses a second downconverter to convert IF signals to baseband [8].

5.2.1 Homodyne Receivers for Vital Sign Detection

Homodyne Doppler radars use a direct conversion receiver to directly convert received signals to the baseband. In a homodyne receiver, the mixer can be a single-channel mixer or a quadrature mixer. A simple homodyne Doppler radar system with a single-channel mixer is shown in Figure 5.3.

The useful vital sign function is included in the phase modulation function in (5.6):

$$\theta(t) = \frac{4\pi}{\lambda} D_v(t) = \frac{4\pi}{\lambda} D_{max} \sin 2\pi f_v t$$

which is linearly proportional to the displacement $D_v(t)$ of the vibrating surface.

For a CW radar, the radar received signal is

$$s_R(t) \cong A_R \cos\left[2\pi f_0 t - \frac{4\pi}{\lambda} R_0 - \theta(t) + \theta_0 + \text{phase noise} \right] \quad (5.10)$$

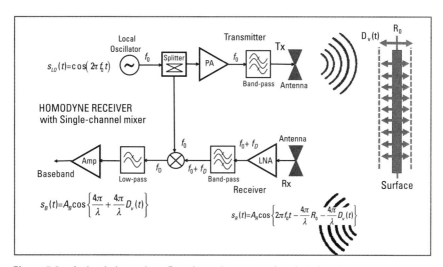

Figure 5.3 A simple homodyne Doppler radar system for vital sign detection.

where R_0 is the distance between the radar and the center of the surface and θ_0 is an initial constant phase shift of the received signal.

After lowpass filtering, the baseband output becomes

$$s_B(t) = A_B \cos\left[\left(\frac{4\pi}{\lambda}R_0 - \theta_0\right) + \frac{4\pi}{\lambda}D_v(t) + n_{\mathrm{Ph}}(t)\right] \qquad (5.11)$$

where $((4\pi/\lambda)R_0 - \theta_0)$ is the constant phase shift dependent on the distance of the target R_0 and a constant phase shift θ_0, and $n_{\mathrm{Ph}}(t)$ is the residual phase noise, that is,

$$n_{\mathrm{Ph}}(t) = \phi_n(t) - \phi_n\left(t - \frac{2R_0}{c}\right) \qquad (5.12)$$

where $\phi_n(t)$ is the phase noise of the oscillator. In CW Doppler radars, the phase noise of the transmitted signal is the noise source in the baseband signals.

If normalizing the amplitude of the baseband signal to $A_B = 1$ and ignoring the phase noise, the baseband signal becomes

$$s_B(t) = \cos\left[\left(\frac{4\pi}{\lambda}R_0 - \theta_0\right) + \frac{4\pi}{\lambda}D_v(t)\right] \qquad (5.13)$$

The baseband signal is a sinusoidal function with a constant phase shift determined by the distance of the target R_0 and the initial phase shift θ_0, and a time-varying phase shift $D_v(t)$ determined by the vibration of the chest wall.

For a human target at a distance R_0 that makes $((4\pi/\lambda)R_0 - \theta_0)$ an integer multiple of π, that is,

$$\left(\frac{4\pi}{\lambda}R_0 - \theta_0\right) = k\pi, \quad (k = 0,1,2,...) \qquad (5.14)$$

such that the baseband signal in (5.13) becomes

$$s_B(t) \cong 1 - \left[\frac{4\pi}{\lambda}D_v(t)\right]^2 \qquad (5.15)$$

Thus, the baseband signal is no longer linearly proportional to the chest wall displacement $D_v(t)$ and, thus, the sensitivity of sensing the vibration is decreased. This is called the null point. There is a null point when the distance R_0 changes every quarter of the wavelength ($\lambda/4$), that is, $R_0 = k(\lambda/4)$, ($k = 0, 1, 2, ...$).

If the distance R_0 makes $((4\pi/\lambda)R_0 - \theta_0)$ to be

$$\left(\frac{4\pi}{\lambda}R_0 - \theta_0\right) = \frac{2k+1}{2}\pi, \quad (k = 0,1,2,...) \tag{5.16}$$

then the baseband signal becomes

$$s_B(t) \cong \frac{4\pi}{\lambda}D_v(t) \tag{5.17}$$

Thus, the baseband signal becomes linearly proportional to the displacement $D_v(t)$, and the sensitivity of sensing the vibration becomes optimal. This is called the optimal point. The null points and optimal points mainly depend on the distance from the radar to the human chest wall, R_0. A null point and the nearest optimal point are only $(\lambda/8)$ apart.

Therefore, a major limitation of the single-channel configuration is that the sensitivity of the detection is related to the distance from the radar to the position of the vibrating surface. However, the null point phenomenon can be avoided by using a quadrature mixer.

5.2.2 Homodyne Receivers with Quadrature Mixer

A homodyne receiver with quadrature-phased dual-channel configuration is shown in Figure 5.4. The architecture consists of two mixers to provide separated baseband outputs to an I-channel and a $\pi/2$-phase-shifted Q-channel [9].

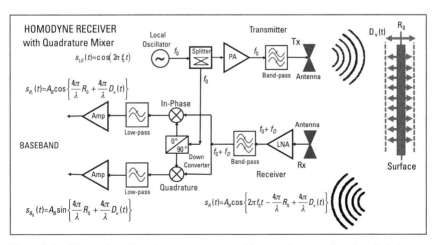

Figure 5.4 A homodyne Doppler radar system with a quadrature mixer for vital sign detection.

With the quadrature dual-channel mixer, the received signal is split into an in-phase channel and a quadrature-phase channel. Ignoring the residual phase noise $n_{Ph}(t)$ and the initial phase shift θ_0, the baseband signal becomes an I-channel baseband

$$s_{B_I}(t) = A_B \cos\left[\frac{4\pi}{\lambda}R_0 + \frac{4\pi}{\lambda}D_v(t)\right] \tag{5.18}$$

and a Q-channel baseband

$$s_{B_Q}(t) = A_B \sin\left[\frac{4\pi}{\lambda}R_0 + \frac{4\pi}{\lambda}D_v(t)\right] \tag{5.19}$$

There are two methods for using the two baseband channels in the quadrature demodulation. One is the normal complex linear quadrature demodulation method and the other is a nonlinear arctangent quadrature demodulation method.

The complex linear quadrature demodulation is shown in Figure 5.5. The quadrature baseband outputs are expressed in (5.18) and (5.19). If the $D_v(t)$ is relatively small and $((4\pi/\lambda)R_0 - \theta_0)$ is an odd multiple of $\pi/2$ in the I-channel signal, or $((4\pi/\lambda)R_0 - \theta_0)$ is an integer number of π in the Q-channel signal, under the condition that the small-angle approximation is applied, the baseband output can be approximated by

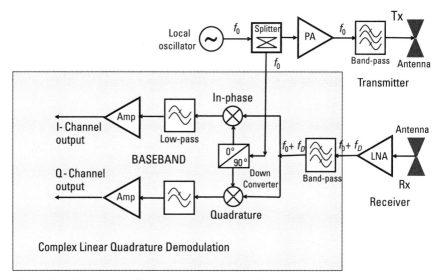

Figure 5.5 The complex linear quadrature demodulation.

$$s_{B_I}(t) = s_{B_Q}(t) \cong \frac{4\pi}{\lambda} D_v(t) \qquad (5.20)$$

Thus, the baseband output is proportional to the chest wall vibration $D_v(t)$.

The nonlinear arctangent quadrature demodulation is shown in Figure 5.6 and the output is given by

$$\arctan\left[\frac{s_{BQ}(t)}{s_{BI}(t)}\right] = \arctan\left\{\frac{\sin\left[\frac{4\pi D_v(t)}{\lambda} + \phi\right]}{\cos\left[\frac{4\pi D_v(t)}{\lambda} + \phi\right]}\right\} = \frac{4\pi}{\lambda} D_v(t) + \phi \qquad (5.21)$$

where $\phi = ((4\pi/\lambda)R_0 - \theta_0)$. The output of the arctangent demodulation is linear proportional to the chest displacement $D_v(t)$ as well as an offset term due to the distance of the chest wall from the radar [10].

If considering direct current (DC) offsets and possible imbalance between the I-channel and the Q-channel, the I-channel and Q-channel baseband signals become

$$s_{B_I}(t) = A_{B_I} \cos\left[\frac{4\pi}{\lambda}R_0 + \frac{4\pi}{\lambda}D_v(t) + \Phi_I\right] + E_{DC_I} \qquad (5.22)$$

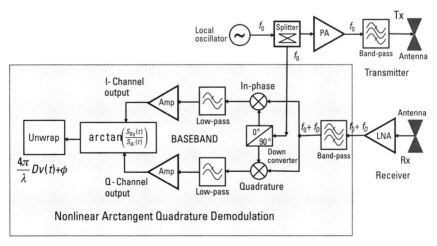

Figure 5.6   The nonlinear arctangent quadrature demodulation.

and

$$s_{B_Q}(t) = A_{B_Q} \sin\left[\frac{4\pi}{\lambda}R_0 + \frac{4\pi}{\lambda}D_v(t) + \Phi_Q\right] + E_{\mathrm{DC}_Q} \qquad (5.23)$$

where (A_{BQ}, Φ_Q) and (A_{BQ}, Φ_Q) are the imbalanced amplitude and phase in the I-channel and the Q-channel, respectively, and $E_{\mathrm{DC}I}$ and $E_{\mathrm{DC}Q}$ are the DC offset in the I-channel and the Q-channel, respectively. Further analysis and calibration of the I-Q imbalance and DC offset can be found in [11].

Although the homodyne receiver does not require an image rejection filter because of the zero-IF, the direct zero-IF conversion has some other disadvantages, such as local oscillator (LO) leakage from the LO port to the mixer input and/or the low noise amplifier (LNA) input that causes self-mixing, the interferer leakage from the LNA input to the LO port, and the $1/f$ (flicker) noise-induced problem. Besides, the quadrature demodulation also faces the I-Q imbalance and DC offset problems [11].

A heterodyne receiver system, which converts the RF signal to an IF signal, can solve the problems caused by the I-Q imbalance and DC offset and avoid problems induced by the flicker noise [12].

5.3 Heterodyne Doppler Radar Systems for Vital Sign Detection

A major problem in vital signs detection is that radar signals modulated by the vibration of human chest wall is very weak and at very low frequency around 0.1 to 3.0 Hz. The baseband signals are superimposed with the $1/f$ flicker noise. It is very difficult to detect a weak signal at a very low frequency and to separate it from a stronger interference signal caused by random body movements or unwanted reflections from static objects.

To overcome the limitation experienced in the direct-conversion homodyne architecture, a heterodyne architecture is often used to improve the receiver's sensitivity significantly. In the heterodyne architecture, the received RF signal is first converted into an intermediate frequency that is well above the flicker noise region. Thus, the flicker noise of the first mixer cannot pass through the IF-bandpass filter. Simultaneously, the IF signal is amplified about 30 to 40 dB in the IF amplifier. Then, in the second mixer, the IF signal is converted into the baseband signal. Because the amplified baseband signal is much stronger than the flicker noise in the second mixing stage, the flicker noise can be ignored.

A simplified heterodyne radar system architecture for vital sign detection is shown in Figure 5.7. In the transmitter, an upconversion mixer is used to mix the oscillator frequency f_1 with the IF f_{IF} to generate a transmit signal at the frequency $f_0 = f_1 \pm f_{IF}$. In the receiver, the first downconversion mixer extracts and amplifies the IF signal. Then, in the second downconversion, the quadrature mixers output the I-channel and Q-channel baseband signals.

5.3.1 Double-Sideband Mixer and Single-Sideband Mixer

In the transmit upconversion mixer, the oscillator frequency f_1 is mixed with an IF f_{IF} to generate an RF transmit signal at the frequency f_0. The transmit signal can be a double-sideband (DSB) signal at $f_1 + f_{IF}$ and $f_1 - f_{IF}$, or a single-sideband (SSB) signal at either $f_1 + f_{IF}$ [13].

If the transmit upconversion mixer is a DSB mixer, the output of the first downconversion mixer in the receiver is also a DSB signal at the IF:

$$s_{DSB}(t) = \cos\left\{\frac{4\pi}{\lambda}D_v(t) + \frac{4\pi}{\lambda}R_0\right\}\cos\left(2\pi f_{IF}t\right) \qquad (5.24)$$

For an SSB transmit upconversion mixer, the receiver IF signal is also an SSB signal at the IF:

$$S_{SSB}(t) = \cos\left\{2\pi f_{IF}t + \frac{4\pi}{\lambda}D_v(t) + \frac{4\pi}{\lambda}R_0\right\} \qquad (5.25)$$

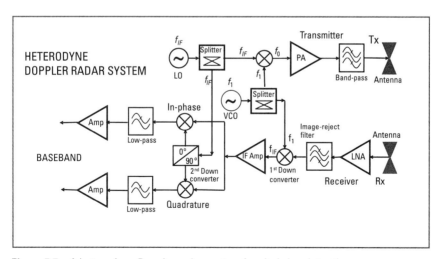

Figure 5.7 A heterodyne Doppler radar system for vital sign detection.

where f_{IF} is the IF, $D_v(t)$ is the chest wall vibration function due to the heart and respiration activity, R_0 is the normal distance to the human chest wall, and λ is the resultant wavelength associated with the upconverted RF signal.

By using a DSB transmit mixer, the IF signal in (5.24) is a sine wave function at the IF with an envelope modulated by the vibration function. Therefore, the problem of null points and optimum points still exists.

However, in the case of an SSB mixer, (5.25) shows that the chest wall modulation is in the phase function of the IF signal. It means that the envelope is no longer modulated by the variation in the distance R_0, such that the problem associated with null points no longer exists. Thus, the SSB mixer in the transmitter is more preferred than the DSB mixer.

5.3.2   The Low-IF Architecture

The low-IF architecture is a type of heterodyne architecture that chooses the IF low enough to be easily digitized at the intermediate frequency [14].

When a signal is digitized at an IF, the DC offsets and the low-frequency noise can be easily removed by using a highpass filter without loss of information. To have a two-channel IF, the receiver should use a quadrature demodulation and, thus, the RF signal is split into two channels. The DC offset can be removed in each of channels by a highpass filter following by a lowpass filter for antialias. Then, signals in each of the I and Q channels are digitized by each of analog-to-digital (A/D) converters, as shown in Figure 5.8.

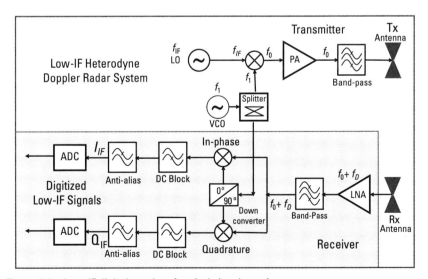

Figure 5.8   Low-IF digital receiver for vital sign detection.

To suppress the phase noise generated in the IF oscillator, the A/D sampling signal should be coherent with the IF signal source.

After digitizing, it is still possible to use an arctangent phase demodulation scheme as a more effective method to remove the DC offsets without any loss of the signal.

Although the low-IF receiver can solve some problems when using the direct-conversion systems, it still needs to be further explored to search more suitable Doppler radar system architectures for the vital sign detection.

5.4 Experimental Doppler Radar for Vital Sign Detection

A Doppler radar with homodyne receiver can be used to study algorithms for radar detection of vital signs. Based on (5.10), the returned RF signal from a vibrating surface can be simplified by

$$s_R(t) \cong A_R \cos\left\{2\pi f_0 t - \frac{4\pi}{\lambda}R_0 - \theta(t)\right\}$$

where R_0 is the distance between the radar and the center position of the surface and the phase modulation function $\theta(t) = (4\pi/\lambda)D_v(t)$ is linearly proportional to the displacement $D_v(t)$ of the vibrating surface.

The displacement of the vibrating chest wall mainly includes two sources: a vibration caused by the heartbeat $D_{\text{Heart}}(t)$ and another vibration caused by the respiratory $D_{\text{Resp}}(t)$:

$$D_v(t) = D_{\text{Heart}}(t) + D_{\text{Resp}}(t) \tag{5.26}$$

From the digitized I-channel $S_{B_I}(t)$ and Q-channel $S_{B_Q}(t)$, the arctangent demodulation can be easily obtained to have the phase modulation function:

$$\arctan\left[\frac{s_{BQ}(t)}{s_{BI}(t)}\right] = \frac{4\pi}{\lambda}\left[D_{\text{Heart}}(t) + D_{\text{Resp}}(t) + R_0 + n_{\text{Ph}}(t)\right] \tag{5.27}$$

which includes the heartbeat, respiration, a constant of the distance R_0 and a phase noise term $n_{\text{Ph}}(t)$.

Because the phase values are between $[-\pi, \pi]$, it needs to be phase-unwrapped to have the true phase function. Then an operation by taking successive phase difference may be applied to the unwrapped phase function to remove unnecessary phase drifts. Other detrending algorithms may also work well to remove the phase drifts. To separate the heartbeat signal from the respiratory signal, two parallel bandpass infinite impulse response (IIR) filtering must be applied. By taking the fast Fourior transform (FFT) the spectrum can be estimated. The detailed processing may also include peak-finding, thresholding, zero-padding, and the estimation of the pulse rate and breathing rate.

Since any Doppler radar system at any frequency band can be used for the detection of vital signs, for health-care monitoring and medical surveillance, the S-band, C-band, X-band, K-band, and even W-band are often used; for through-the-wall vital sign detection and for finding survivors after an earthquake, the S-band and C-band are used. The simplest radar system for the detection of vital signs is the monostatic CW or FMCW radar systems with homodyne receiver. An experimental X-band CW radar for monitoring a stream of live human heartbeat and breathing data is illustrated in Figure 5.9. From the continuous flow of data, the respiratory function and heartbeat function can be clearly monitored in real time.

Based on data processing algorithms in Sections 5.2.1 and 5.2.2, vital signs can be extracted from radar collected data. A radar used for the vital

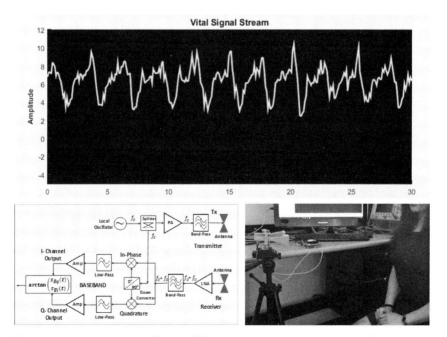

Figure 5.9 An experimental X-band CW radar used to monitor a stream of human heartbeat and breathing data.

sign data collection is a K-band radar operating at 25 GHz with a CW waveform, 128-kHz data sampling rate, and 10-second recording time for each data collection.

The data processing procedure for CW signals is straightforward. After signal conditioning, such as DC cancellation, data rate reduction, and I-Q imbalance correction, the arctangent demodulation operation, phase unwrapping, differencing, and detrending processing are applied to the digitized baseband signals. The estimated spectrum of the heartbeat and respiratory signals using the K-band radar is shown in Figure 5.10.

The FMCW waveform is also often used for vital sign detection to provide additional target range information. After DC cancellation and I-Q imbalance correction, by taking FFT along the fast-time, a two-dimensional (2-D) range profiles in the range and slow-time domain can be formed as shown in Figure 5.11(a). From the peak of the averaging range profile, the target's range can be estimated and, thus, a range gate can be determined. By averaging data over the range gate, preconditioned I-channel and Q-channel signals are ready for further processing. Then, following the data processing procedure for CW signals, such as the arctangent demodulation operation, unwrap, differencing, and detrending processing, after two-bandpass filtering, the estimated spectra of the heartbeat signals and the respiratory signals are shown in Figure 5.11(b, c).

For improving the radar performance for vital sign detection, the sensitivity of the radar receiver must be enhanced and the flicker noise must be suppressed. Thus, the heterodyne architecture is preferred.

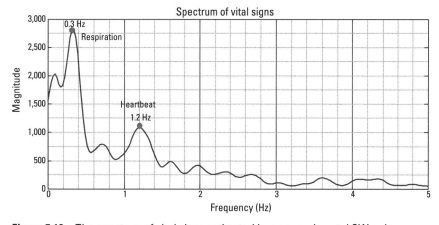

Figure 5.10 The spectrum of vital signs estimated by an experimental CW radar.

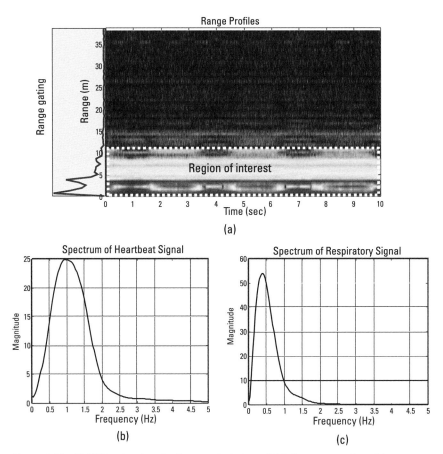

Figure 5.11   FMCW radar data for human vital signs: (a) 2-D range profiles, (b) spectrum of heartbeat signals, and (c) spectrum of respiratory signals.

A set of collected homodyne Doppler radar data for vital sign study is provided in this book. Both CW radar data and FMCW radar data are collected. A set of MATLAB codes for processing vital sign data are provided.

References

[1]   Chen, K. M., and H. -R. Chuang, "Measurement of Heart and Breathing Signals of Human Subjects Through Barriers with Microwave Life-Detection Systems," *Proceedings of Annual International Conference of the IEEE Engineering in Medicine and Biology Society*, New Orleans, LA, 1988, pp. 1279–1280.

[2] Narayanan, R. M., "Earthquake Survivor Detection Using Life Signals from Radar Micro-Doppler," *Proceedings of the 1st International Conference on Wireless Technologies for Humanitarian Relief*, 2011, pp. 259–264.

[3] Gu, C. Z., "Continuous-Wave Radar Sensor Based on Doppler Phase Modulation Effect for Medical Applications and Mechanical Vibration Monitoring," Ph.D. thesis, Texas Tech University, 2013.

[4] Lin, J. C., "Noninvasive Microwave Measurement of Respiration," *Proceedings of the IEEE*, Vol. 63, No. 10, 1975, pp. 1530–1530.

[5] Chen, K. M., et al., "An X-Band Microwave Life-Detection System," *IEEE Transactions on Biomedical Engineering*, Vol. 33, No. 7, 1986, pp. 697–702.

[6] Chen, K. M., et al., "Microwave Life-Detection Systems for Searching Human Subjects Under Earthquake Rubble or Behind Barrier," *IEEE Transactions on Biomedical Engineering*, Vol. 27, No. 1, 2000, pp. 105–114.

[7] Razavi, B., "Design Considerations for Direct-Conversion Receivers," *IEEE Transactions on Circuits and Systems—II: Analog and Digital Signal Processing*, Vol. 44, No. 6, 1997, pp. 428–435.

[8] Gruz, P., H. Gomes, and N. Carvalho, "Receiver Front-End Architectures—Analysis and Evaluation," in *Advanced Microwave and Millimeter Wave Technologies Semiconductor Devices Circuits and Systems*, M. Mukherjee (ed.), InTech, 2010.

[9] Raffo, A., S. Costanzo, and V. Cioffi, "Quadrature Receiver Benefits in CW Doppler Radar Sensors for Vibrations Detection," in A. Rocha, et al. (eds.), *Trends and Advances in Information Systems and Technologies, WorldCIST'18 2018, Advances in Intelligent Systems and Computing*, Vol. 746, 2018, pp. 1471–1477.

[10] Park, B. -K., O. Boric-Lubecke, and V. M. Lubecke, "Arctangent Demodulation with DC Offset Compensation in Quadrature Doppler Radar Receiver Systems," *IEEE Transactions on Microwave Theory and Techniques*, Vol. 55, No. 5, 2007, pp. 1073–1079.

[11] Park, B. -K., S. Yamada, and V. Lubecke, "Measurement Method for Imbalance Factors in Direct-Conversion Quadrature Radar Systems," *IEEE Microwave and Wireless Components Letters*, Vol. 17, No. 5, 2007, pp. 403–405.

[12] Churchill, F. E., G. W. Ogar, and B. J. Thompson, "The Correction of I and Q Errors in a Coherent Processor," *IEEE Transactions on Aerospace and Electronic Systems*, Vol. 17, No. 1, 1981, pp. 131–137.

[13] Jensen, B. S., et al., "Vital Signs Detection Radar Using Low Intermediate Frequency Architecture and Single-Sideband Transmission," *The European Microwave Conference*, Amsterdam, The Netherlands: RAI, 2012.

[14] Jensen, B. S., T. K. Johansen, and L. Yan, "An Experimental Vital Signs Detection Radar Using Low-IF Heterodyne Architecture and Single-Sideband Transmission," *Proceedings of 2013 IEEE International Wireless Symposium (IWS)*, 2013.

6

Application to Hand Gesture Recognition

Gestures and postures are both important forms of human body language. The difference between postures and gestures is that a posture refers to a static configuration of the position and orientation of specific human body parts, but a gesture refers to dynamic characteristics or a sequence of postures presented within a specific time period.

Although a posture is sometimes called a gesture as in sign language [1], strictly speaking, however, a gesture should be a meaningful physical movement of fingers, hands, arms, or other parts of the human body, with the purpose to convey information or meaning for environmental interactions.

Therefore, a static pose of fingers and hands, such as the victory sign, the okay sign, or the pause sign, is called a hand posture. A hand movement involving fingers, a palm, and a wrist with the intent of conveying meaningful information and interacting with the environment, such as waving hands for good-bye, rubbing the thumb against the fingertips for a money expectancy, or a finger slid across the throat for strong disapproval, is called a hand gesture.

Hand gestures can be interpreted as semantically meaningful commands and can be applied to human-computer interaction, navigating, and manipulating in virtual environments, automated homes and offices, robot control, and so forth.

When fingers and hands are moving in different directions and at different speeds, they can cause Doppler frequency shifts in radar and appear with different shapes, intensities, and durations in radar micro-Doppler signatures.

The micro-Doppler signature of a hand gesture is a characteristic of hand movement that represents the intricate frequency modulation generated by the movement of the palm, wrist, and fingers. It is a principal kinematic feature of the hand movement for hand gesture recognition. Together with other features, such as distances, angular velocities, or angles of arrival, more complicated or fine-grained hand gestures can be recognized in the complex feature space. To improve the performance of gesture recognition, correct interpretation of principal features and establishment of a mapping between features and hand gestures are needed. This process allows one to extract relevant quantitative and objective features.

The entire gesture recognition process involves motion capture, feature extraction, pattern recognition, and machine learning. The purpose of this chapter is to review and discuss the role of radar micro-Doppler signatures as an important feature played in hand gesture recognition. The modeling and capturing of micromotions of the hand and fingers are reviewed in Sections 6.1 and 6.2. Section 6.3 describes radar micro-Doppler signatures for hand gesture recognition. Other features for hand gesture recognition are discussed in Section 6.4.

Artificial neural networks (ANNs) are one of the advanced technologies that solve the pattern recognition task as an artificial brain in machine learning. Deep learning is a subset of machine learning that can create an ANN as an artificial brain and perform any recognition task directly from images, video, audio, and even text.

Convolutional neural networks (CNNs) are one of the most popular algorithms for deep learning without the need for manual feature extraction and are particularly useful for recognizing objects, faces, and scenes.

However, the recognition task using CNNs and deep learning is beyond the scope of this book. Many publications [2–7] and the MATLAB toolbox on ANNs, CNNs, and deep learning are available.

6.1 Modeling of Hand and Finger Movement

The human hand has a very complex structure that consists of 27 bones with interlinking joints. Each hand has 5 fingers (thumb, index, middle, ring, and little finger), and each finger consists of 3 joints. The little, ring, middle, and index fingers are aligned together and connected to the wrist bones. The thumb is on the side of the 4 fingers for capturing, grasping, and holding. Movements of human hand joints, such as deviation, twist, flexion or extension, abduction, or adduction, depend on the type of movement and the possible rotation axes.

Kinematic parameters of human hand movements include position, velocity, acceleration, and angular orientation. To completely describe any activity of a hand gesture in a three-dimensional (3-D) Cartesian coordinate system, the linear kinematic parameters (position, velocity, and acceleration) define the manner of changes of any point in the hand and fingers over time. The angular orientation or the joint angle of a finger segment is also an important kinematic parameter. Together with the angular velocity and acceleration, the three angular kinematic parameters completely describe angular motions of the hand and fingers.

The kinematic model of human hands depends on the users and applications. The computer vision community often uses a kinematic model with 27 degrees of freedom (DOFs), including 5 DOFs for the thumb, 16 DOFs for the other four fingers, 3 DOFs for the wrist rotation, and 3 DOFs for the wrist translation. If not considering the wrist translation, the model with 24 DOFs may be sufficient for the gesture recognition.

A 3-D model of human hand and wrist is shown in Figure 6.1, where the hand has 27 bones with 24 DOFs for rotation in a 3-D space.

6.2 Capturing of Hand and Finger Movements

There are various hand motion capture systems, such as optical cameras, infrared (IR) cameras, special gloves and markers, and markerless motion capture systems.

6.2.1 Traditional Motion Capture Methods

Traditional motion capture methods for hand gesture include: (1) vision-based methods, (2) glove-based methods, and (3) marker-based motion capture methods.

The vision-based method uses one or more cameras to collect two-dimensional (2-D) spatial image sequence. A 3-D time-of-flight (TOF) camera can capture the range image to provide the depths of scene points. The TOF sensor measures the depths in real time by illuminating the scene with a laser, infrared, or millimeter-wave source and thus captures full spatial and temporal 3-D scenes [8].

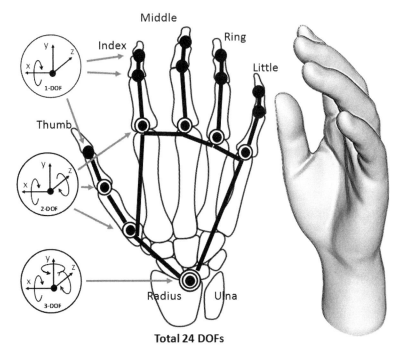

Total 24 DOFs

Figure 6.1   A 3-D model of the hand and wrist.

An example of such a 3-D TOF camera is a tiny camera proposed in [9], which uses an infrared laser to create an invisible light for illuminating the surrounding environment and uses a wide-angle IR camera to sense the reflected light from the environment for measuring the distance information about the environment.

The glove-based method uses input devices on an instrumented glove worn by the user. A marker-based optical motion capture system has the capability of tracking 3-D positioning. The system is equipped with number of infrared cameras to record 3-D trajectories of hand gestures collected by markers placed on the hand [10]. Due to too many DOFs of joints distributed on a smaller hand, the markers must be very small and placed close to each other. Each finger may need 2 or 3 markers and a total of as many as 27 markers may be needed on one hand [11].

The marker-based method can accurately track hand and finger movements. Even if for very complex gestures, such as pianists' hand and fingers movements, current technology can track very accurately as reported in [12], in which 12 infrared cameras were used with and 27 markers were placed on each hand to allow accurate capturing all DOF and recording 3-D motion trajectories of the fingers, palm, and wrist with a frame rate of 100 frames/second. An example of 27-marker configuration is shown in Figure 6.2(a) and an example of the captured pianists' hands and fingers is shown in Figure 6.2(b).

However, visible and infrared cameras have some drawbacks. They are expensive, requiring a huge memory, computing power, and communication bandwidth to process and transmit the images, are relatively immobile, and raise privacy concerns when deploying the system in the public. Therefore, acoustic and radar-based systems are good alternatives to visible and infrared camera-based systems for the sense of hand gestures.

6.2.2 Acoustic Doppler-Based Systems for Hand Gesture Recognition

Acoustic (or ultrasonic) Doppler-based hand gesture recognition is contactless and markerless. This is an alternative to systems using cameras, gloves, and markers. The acoustic system does not invade personal privacy and works well at night or in smoke-filled areas with poor visibility [13–18].

The transmitter is a loudspeaker that generates acoustic or ultrasonic waves. For the sensing of hand gestures, the acoustic signal must be inaudible to humans. Thus, the acoustic frequency is usually near the upper end of the human hearing frequency range: 18–19 kHz. The acoustic receiver is a microphone. Multiple microphones may be used to receive Doppler-shifted acoustic waves. The general 44.1-kHz sampling rate of the typical microphone can be used to digitize Doppler-shifted acoustic signals.

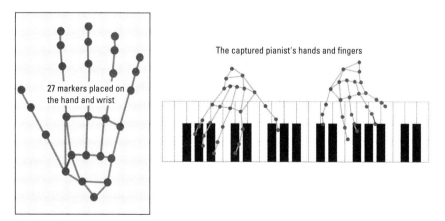

Figure 6.2 (a) Marker configuration with 27 markers placed on the hand and wrist, and (b) an example of the captured pianists' hands and fingers.

The SoundWave system [13] uses one loudspeaker and one microphone. It can correctly recognize a set of five one-dimensional (1-D) gestures. The AudioGest system [14] also uses a pair of loudspeakers and a microphone to sense fine-grained hand movements and be able to detect six hand gestures with high accuracy. Both the SoundWave and the AudioGest systems use rule-based heuristics to recognize small sets of gestures without using training data.

To recognize larger number of gestures in more complex environments, a system called the MultiWave was proposed. It can support 14 classes of gestures using only a few training samples [15].

A Doppler sonar at 40-kHz ultrasound frequency band for the recognition of 3-D single handed gestures was introduced in [16]. The Doppler sonar consists of one ultrasonic transmitter and three ultrasonic receivers, which can recognize eight hand gestures, including directional swiping in 3-D and rotational gestures.

For tracking fine-grained finger movements, a multi-microphone acoustic system named FingerIO was proposed [17]. It transmits an inaudible 18–20-kHz sound wave and receives from multichannel microphones. The system can measure the location of each finger from its distances to microphones. Orthogonal frequency division multiplexing (OFDM) is utilized to track fingers and achieve average 2-D tracking accuracies of 8–12 mm.

However, acoustic and ultrasonic sensors have their own disadvantages. They are easy to be interfered with by environmental sound sources and have poor penetration properties through materials. Thus, radar-based Doppler systems have an advantage in sensing dynamic hand gestures.

6.2.3 Radar Doppler-Based Systems for Hand Gesture Recognition

The Doppler radar sensor is highly sensitive to micromovements and has excellent ability to distinguish moving objects from stationary clutter and background. It does not require gloves and markers attached to the human hand and can easily penetrate plastic and some other materials. Thus, the Doppler radar sensor has potential to be a better sensor for hand gestures.

In recent years, Doppler radar systems, especially continuous-wave (CW) and frequency-modulated continuous wave (FMCW) radars, are being increasingly utilized in hand gesture recognition and have achieved competitive results. Many advanced technologies in feature extraction, decomposition, machine learning, and ANNs have been successfully applied to radar hand gesture recognition [19–24].

The mechanical movements of hands and fingers cause phase changes in the reflected radar signals. Based on the phase changes, it is possible to measure radial velocities and trajectories of hands and fingers.

Based on micro-Doppler signatures, range-Doppler maps, 3-D motion trajectories, and other possible features, Doppler radar systems have been successfully applied to fine-grained hand gesture recognition and control [21–24].

Depending on the operating frequency band and signal bandwidth, most Doppler radars can only capture relatively large movements of hands and fingers, such as waving hand. In some applications, it is necessary to recognize fine-grained or micro hand gestures with just a few fingers. To differentiate a subtle change in position, the required range resolution $\Delta r = c/(2B)$ should be better than 1–2 cm and the velocity resolution $\Delta v_r = c/(2f_cT)$ should be better than 0.1 cm/s, where c is the speed of the wave propagation, B is the signal bandwidth, T is the signal integration time, and f_c is the transmit carrier frequency. Thus, for capturing fine-grained hand gestures, the radar transmit frequency must be high enough and the signal bandwidth must be wide enough. The radar gesture sensor demonstrated by the Google's Project Soli [25] is a millimeter-wave FMCW radar operating at 60 GHz with 7 GHz of bandwidth for best control smart devices through fine-grained finger motions [24]. In addition to micro-Doppler signatures, the Soli radar also utilizes other available features, such as 1-D range profile, 1-D Doppler profile, 2-D range profiles, and 3-D time-varying range-Doppler maps.

6.3   Radar Micro-Doppler Signatures for Hand Gesture Recognition

Micro-Doppler signature is a principal feature of micromotions. Based on only the micro-Doppler feature without using information gained from other features (such as range profiles, Doppler profiles, range-Doppler maps, or time-varying range-Doppler maps), it is also possible to realize simple hand gesture recognition and control.

The micro-Doppler signature of a hand gesture is a kinematic property of the hand movement that represents the intricate Doppler frequency modulations by the movement. Radar micro-Doppler signatures are insensitive to the distance, lighting condition, and background complexity. To recognize a hand gesture, the radar-captured data must be processed and decomposed into a feature space. Then the CNNs and deep learning may be applied to recognize the hand gestures [19, 20, 24, 26].

To improve the performance of the gesture recognition, correct interpretation of the micro-Doppler signature and establishment of a mapping between the signature and the hand gesture are needed. This process allows relevant quantitative and objective features to be extracted for the comparison between different gestures.

To interpret the Doppler shifts of a hand gesture, it is helpful to show the relation between the micro-Doppler signature and the hand gestures. Since the speed of the wave propagation c is much greater than the radial velocity of a finger's motion, v_{radial}, the Doppler frequency shift is

$$f_D = -\frac{2f_c}{c}v_{radial} \tag{6.1}$$

If the path of the finger motion has an angle θ with respect to the receiver and the speed is v_{finger}, the radial velocity of the moving finger becomes

$$v_{radial} = v_{finger}\cos\theta \tag{6.2}$$

and, thus,

$$f_D = -\frac{2f_c}{c}v_{finger}\cos\theta \tag{6.3}$$

To interpret how to link a gesture motion with a micro-Doppler signature, Figure 6.3 depicts a gesture of snapping fingers. When the middle finger is moving at a velocity, v_{Middle}, from the upper left to stop at the lower right, the angle θ_{Middle} gradually decreases from about $\pi/6$ to about $\pi/9$; hence, $\cos(\theta_{Middle})$ increases from 0.87 to 0.94. As a result, the Doppler frequency shift increases from $-0.87v_{Middle}(2f_c/c)$ to $-0.94v_{Middle}(2f_c/c)$. At the same time, the thumb is moving at the velocity, v_{Thumb}, from the lower right to stop at the upper left, the angle θ_{Thumb} gradually increases from about $\pi/4$ to about $\pi/3$; hence, $\cos(\theta_{Thumb})$ decreases from 0.7 to 0.5. Thus, the Doppler frequency shift decreases from $0.7v_{Thumb}(2f_c/c)$ to $0.5v_{Thumb}(2f_c/c)$.

The micro-Doppler signature of the snapping fingers in Figure 6.3 indicates that there is a higher negative Doppler peak and a lower positive Doppler peak located at the same time instant. The higher negative Doppler peak corresponds to flipping the middle finger onto the palm and the lower positive Doppler peak corresponds to flipping the thumb toward the radar. Then the three fingers are reset back to their initial positions and there is another low positive Doppler peak appearing as shown in the figure.

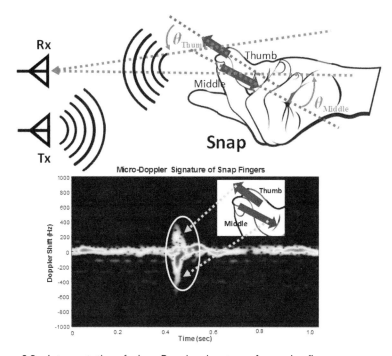

Figure 6.3 Interpretation of micro-Doppler signature of snapping fingers.

To model the time-varying radial velocity, the Doppler frequency shift is related the radial velocity by

$$f_D(t) = -\frac{2f_c}{c} v_{\text{radial}}(t) = -\frac{2}{\lambda} v_{\text{radial}}(t) \tag{6.4}$$

and, thus, the time-varying radial velocity is

$$v_{\text{radial}}(t) = -\frac{\lambda}{2} f_D(t) \tag{6.5}$$

where λ is the wavelength of the transmit signal.

At each time instant, the micro-Doppler frequency shift is $f_D(t)$ and the radial velocity of the finger $v_{\text{radial}}(t)$ is related by a half wavelength $\lambda/2$. The sign of the radial velocity indicates the direction of motion, which relates to the type of gesture. The time interval of nonzero velocity represents the duration of the gesture. Also, based on the area covered by the radial velocity curve, the distance of the finger from the radar can be estimated.

Based on radar micro-Doppler signatures of a set of preselected hand gestures, simple monostatic CW Doppler radar can perform very well for the hand gesture recognition. An example of a micro-Doppler signature-based hand gesture recognition system was demonstrated for controlling a CD player [26].

For simplicity, the demonstration only used four gestures to control a CD player: snap fingers for "Play," swipe away for "the Next," swipe toward for "the Previous," and flip fingers for "Stop," as illustrated in Figure 6.4.

The corresponding four micro-Doppler signatures for controlling a CD player are shown in Figure 6.5. With a simple monostatic CW radar and only based on micro-Doppler features, the radar sensor can reach an accuracy rate of classification higher than 88%.

Although it is simple and natural to utilize micro-Doppler signatures for hand gesture recognition, the Doppler effect only reflects the radial velocity of a moving object. When the moving object has no radial velocity components, there will be no Doppler shift. If an object moves along a curved path, when its radial velocity decreases, the angular velocity must increase. Thus, to completely describe the object's motion, both its radial and angular velocities are needed. Section 1.10 in Chapter 1 introduced the interferometric frequency

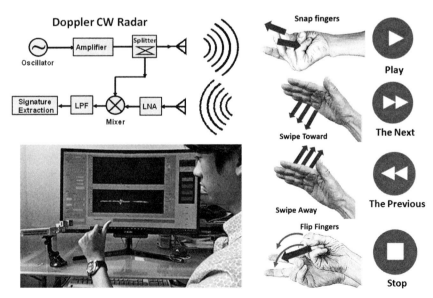

Figure 6.4 An example of micro-Doppler signatures for hand gesture recognition and control.

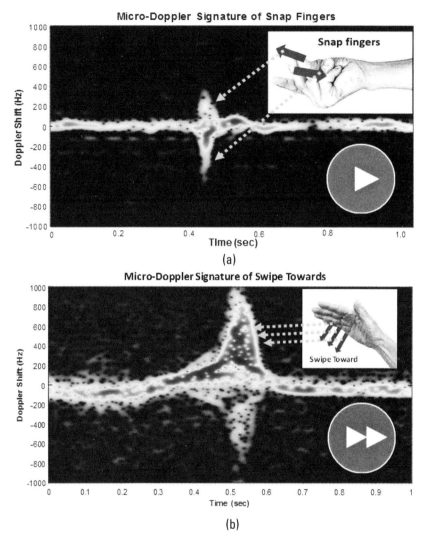

Figure 6.5 (a) The micro-Doppler signature for "Play," (b) the micro-Doppler signature for "the Next," (c) the micro-Doppler signature for "the Previous," and, (d) the micro-Doppler signature for "Stop."

shift f_{Inf}, which is proportional to the angular velocity, while the Doppler frequency shift f_D is proportional to the radial velocity.

As well as the feature from micro-Doppler signatures, additional features such as time-varying range-Doppler maps, azimuth and elevation angles, and

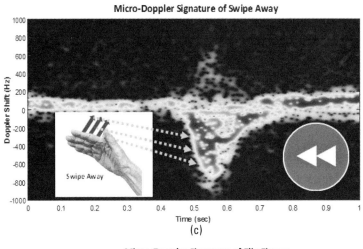

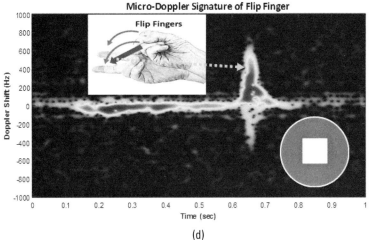

Figure 6.5   *(Continued)*

3-D trajectories can also be utilized to enhance the performance of the hand gesture recognition.

6.4   Other Features for Hand Gesture Recognition

To sense precise and subtle micromotions for fine-grained hand gesture recognition and to enhance the performance of the hand gesture recognition, more other features are often needed.

6.4.1 Time-Varying Range-Doppler Features

In a wideband Doppler radar, a range-Doppler map can be easily formed [27]. The range-Doppler map is a 2-D complex (amplitude and phase) function distributed in the joint range and Doppler domain. Each moving object appears as a peak on the range-Doppler map. During a certain signal integration time duration, a sequence of range-Doppler frame forms a 3-D time-varying range-Doppler maps. The peaks appearing within the 3-D maps formulate the movement of the object of interest.

For an FMCW radar, given the number of range samples, M, in each sweep time, after taking the FFT along the M range cell samples, one range profile can be formed. For a group number of N sweeps (also called the number of slow-time samples), after taking another FFT along the N sweeps, a Doppler profile is formed. Then the combined range profiles and Doppler profiles form an $M \times N$ range-Doppler map. By repeating the above processing for the entire signal integration time duration, a sequence of $M \times N$ range-Doppler map frames is generated. Every range-Doppler map frame is at a specific time t. A 3-D time-varying range-Doppler map is shown in Figure 6.6.

Thus, the hand gesture data becomes a 3-D range-Doppler-time cube, which is the stack of range-Doppler maps among the time dimensions and provides time-varying distances and radial velocities (i.e., micro-Doppler

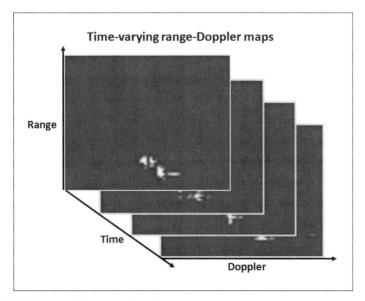

Figure 6.6 Time-varying range-Doppler map.

signature) of the hand, palm, and fingers. By taking a slice at a range cell, it can show the micro-Doppler signature of the palm of the hand and fingers at that range cell.

6.4.2 Azimuth and Elevation Angle Features

Based on monopulse principles [28], a four-channel Doppler radar system with two pairs of collocated horizontal and vertical receiving antennas can measure azimuth and elevation angles as illustrated in Figure 6.7.

The angle of arrival of the echo from an object is estimated by comparing signals received at two collocated receiver pairs. The azimuth angle of arrival is estimated by

$$\varphi_{az} = \arcsin\left(\frac{\lambda\Delta\psi_{12}}{2\pi D}\right) \tag{6.6}$$

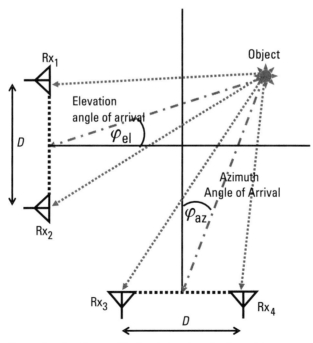

Figure 6.7 Two pairs of collocated horizontal and vertical receiving antennas for measuring the azimuth and elevation angles of arrival.

where D is the geometrical distance between two azimuth or elevation antennas, λ represents the wavelength, and $\Delta\psi_{12}$ is the observed phase difference between the two signals in receivers 1 and 2. Similarly, the elevation angle of arrival is estimated by

$$\varphi_{el} = \arcsin\left(\frac{\lambda\Delta\psi_{34}}{2\pi D}\right) \qquad (6.7)$$

where $\Delta\psi_{34}$ is the observed phase difference between the two signals in receivers 3 and 4.

The range information is gained by using a wideband signal, such as FMCW. Therefore, monopulse radars can measure the range, velocity, azimuth, and elevation angles of moving objects in the scene. From radar received data, a sequence of 2-D complex range-Doppler maps can be generated in each receiving channel. Thus, the angle of arrival can be estimated simultaneously from the phase differences between two range-Doppler maps of each pair as proposed in [21].

In the range-Doppler map, moving objects appear as peaks. Multiple objects may be overlapped or spatially separated on the map. The phase difference, $\Delta\psi_{12}(r, f_D)$, between the two range-Doppler maps in receivers 1 and 2 is used to estimate the azimuth angles of moving objects. The phase difference, $\Delta\psi_{34}(r, f_D)$, between the two range-Doppler maps in receivers 3 and 4 is for the elevation angles, where r is the range and f_D is the Doppler shift of each detected moving object. Thus, a four-receiver FMCW radar is needed to measure the range, velocity, azimuth, and elevation angles of moving objects in the scene.

From radar reflected signals, by comparing the phases for each pairs of range-Doppler maps, the 3-D spatial coordinates in range, azimuth, and elevation and corresponding velocities can be estimated for the hands and fingers. A unique feature of time-varying range-Doppler-angle representation can be utilized in the hand gesture recognition.

6.4.3 Fine-Grained Hand Gesture Recognition

The sensing of precise and subtle micro motions, such as sliding the thumb over the index finger, can be used in fine-grain hand gesture control, such as adjusting the dial on a watch. To extract and recognize subtle motions from radar signals, this is a challenge.

6.4.3.1 High-Resolution Spatial and Temporal Processing

For the sensing of fine-grained hand gestures, radar sensors would be better to work at a millimeter-wave band, such as W-band, and with an ultrawide bandwidth. The range resolution is determined by $c/(2B)$, where c is the speed of wave propagation and B is the signal bandwidth. For example, for a signal with 7-GHz bandwidth, the radar can reach a range resolution of 2.14 cm. The velocity resolution is determined by $c/(2f_cT)$, where f_c is the transmit carrier frequency and T is the signal integration time. For a 60-GHz carrier frequency and 32-ms signal integration time, the velocity resolution can be 7.81 cm/s. Thus, having a high-resolution range-Doppler map, closely distributed dominant scatterer centers can be resolved by their range cells and/or Doppler cells.

The temporal processing also helps for resolving features. An unresolvable shape of an object in the spatial domain may be resolved in the temporal domain. Therefore, a sequence of time-varying 2-D range-Doppler maps may clearly display dynamic hand gesture patterns over time [23, 24].

6.4.3.2 Multi-Input-Multi-Output Radar for Tracking of 3-D Trajectory

Google Soli radar uses two transmit and four receive antenna elements through digital beamforming for azimuth and elevation angular tracking [23, 24]. Multi-input-multi-output (MIMO) radar is an advanced technology for tracking in 3-D spatial coordinates and becomes an extension to the digital beamforming radar [29]. The combination of phased array radar with the MIMO concept makes the MIMO phased array radar an attractive technology, especially for short-range radar applications.

The main difference between MIMO radars and conventional radars is that the MIMO radar is capable of transmitting different signals from different transmit antennas and keeping these signals separable at receivers. MIMO technology can synthesize virtual antenna positions that lead to large numbers of effective array elements and gain a higher spatial resolution.

The MIMO radar consists of multiple antenna elements in both the transmission side and the reception side. For N transmit elements and M receive elements, there are $N \times M$ distinct propagation channels from the transmit array to the receive array. A virtual phased array of $N \times M$ elements can be synthesized with only $N + M$ antenna elements. The receive antennas must be able to separate the signals corresponding to different transmit antennas. The diversity of transmit channels can be achieved by using time division multiplexing, frequency division multiplexing, spatial coding, or orthogonal waveforms. The virtually formed phased array can be designed to produce the desired pattern by arranging the placements of the transmit elements and receive elements in an appropriate way. By using a convolution of phase center positions, a 2-D MIMO virtual array can be formed as a desired pattern. The MIMO virtual array is equivalent to single transmit phase center with a 2-D virtual receive array, as illustrated in Figure 6.8. Because N transmit elements and M receive elements can form a virtual phased array of $N \times M$ elements, with 2 transmit antenna elements and 4 receive elements, the virtual receive array has 8 elements as illustrated in the figure. The desired pattern of the virtual receive array can be managed by selecting appropriate placement of the transmit elements or the receive elements.

By arranging the placements of the transmit and receive elements in a different way, a larger-aperture sparse virtual receive array can be formed, and a higher spatial resolution can be achieved.

6.4.4 Radar Frontal Imaging of Hand Gestures

A synthetic aperture radar (SAR) image is a top-view image displayed in the range and cross-range dimension that may not capture much information about the movement of the object of interest. The motion-induced phase error causes the SAR image of the object to be mislocated in the cross-range dimension and smeared in both the cross-range and the range domains. Moving objects appear as defocused and spatially displaced, superimposed on the SAR map. Thus, SAR is impractical for realizing radar images of moving objects.

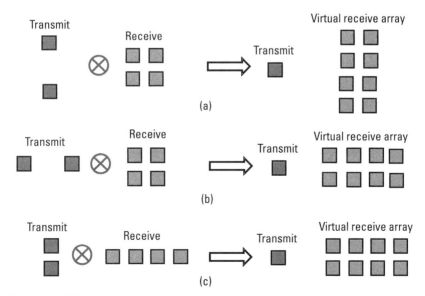

Figure 6.8 With two transmit antenna elements and four receive elements, a 2-D virtual receive array pattern with eight elements can be formed.

Compared to the top-view SAR images, a more informative viewing of an object's activities is the frontal view. Instead of using synthetic aperture, a larger 2-D virtual array is used in the frontal imaging. Thus, a time-varying Doppler-azimuth-elevation image sequence of an object's movement can be reconstructed in a 3-D Fourier space. In [30], a CW Doppler radar operating at 7.5 GHz is assumed. The antenna is a uniform planar antenna aperture with 20×20 half-wavelength spaced elements. The radar data is Fourier transformed across 3-D Doppler, azimuth, and elevation domain. Thus, an image can be generated by a complex sum of the 2-D cross-range point spread responses of each distinct scatterer in the 3-D Fourier space. Although the number of antenna elements is 400, the arms and legs of the human can still be distinguished.

A method for micro-Doppler enhanced frontal radar imaging of human activities was described in [31, 32]. The micro-Doppler signatures arising from the dynamic movements of the arms and legs are used to resolve some of the point scatterers on the human body. The additional Doppler dimension enables the relaxation of the resolution criteria across the cross-range dimensions.

The main advantage of micro-Doppler enhanced frontal imaging is that the radar operators may be able to directly infer a wide variety of human activities without the assistance of complicated machine learning algorithms. Figure 6.9 shows one slice from a sequence of frontal images generated by simulation of a human skipping. For easy comparison, the ground truth image is displayed

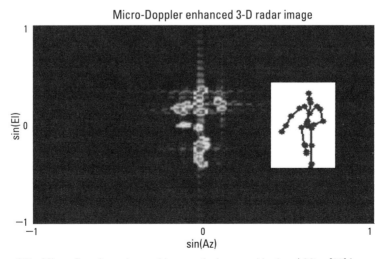

Figure 6.9 Micro-Doppler enhanced image of a human skipping. (*After:* [32].)

on the right. The micro-Doppler enhanced frontal image shows both arms, the torso, the head, and one leg. The most interesting feature of these images is that they capture the bobbing motion of the head during skipping. However, the micro-Doppler enhanced frontal imaging has a limitation. It cannot be used to image stationary objects.

References

[1]   Mitra, S., and T. Acharya, "Gesture Recognition: A Survey," *IEEE Transactions on Systems, Man and Cybernetics, Part C: Applications and Reviews*, Vol. 37, No. 3, 2007, pp. 311–324.

[2]   Samuel, A., "Some Studies in Machine Learning Using the Game of Checkers," *IBM Journal of Research and Development*, Vol. 3, No. 3, 1959, pp. 210–229.

[3]   LeCun, Y., and Y. Bengio, "Convolutional Networks for Images, Speech, and Time-Series," in M. A. Arbib, (ed.), *The Handbook of Brain Theory and Neural Networks*, Cambridge, MA: MIT Press, 1995.

[4]   LeCun, Y., Y. Bengio, and G. Hinton, "Deep Learning," *Nature*, Vol. 521, May 2015, pp. 436–444.

[5]   Schmidhuber, J., "Deep Learning in Neural Networks: An Overview," *Neural Networks*, Vol. 61, No. 1, 2015, pp. 85–117.

[6]   Molchanov, P., et al., "Hand Gesture Recognition with 3D Convolutional Neural Networks," *2015 IEEE Computer Society Conference on Computer Vision and Pattern Recognition Workshop*, 2015, pp. 1–7.

[7]   Kim, Y., and T. Moon, "Human Detection and Activity Classification Based on Micro-Doppler Signatures Using Deep Convolutional Neural Networks," *IEEE Geoscience and Remote Sensing Letters*, Vol. 13, No. 1, 2016, pp. 8–12.

[8]   Kollorz, E., J. Hornegger, and A. Barke, "Gesture Recognition with a Time-of-Fight Camera, Dynamic 3D Imaging," *International Journal of Intelligent Systems Technologies and Applications*, Vol. 5, No. 3-4, 2008, pp. 334–343.

[9]   "A Hand-Tracking Sensor for Virtual Reality Headsets, the 2015 Invention Awards," *Popular Science*, May 25, 2015.

[10]   Wheatland, N., et al., "State of the Art in Hand and Finger Modeling and Animation," *Computer Graphics Forum*, Vol. 34, No. 2, May 2015, pp. 735–760.

[11]   Kitagawa, M., and Windsor B, *MoCap for Artists: Workflow and Techniques for Motion Capture*, Burlington, MA: Focal Press, 2008.

[12]   Tits, M., et al., "Feature Extraction and Expertise Analysis of Pianists' Motion-Captured Finger Gestures," *ICMC*, 2015. September 25–October 1, 2015, CEMI, University of North Texas.

[13] Gupta, S., et al., "SoundWave: Using the Doppler Effect to Sense Gestures," *ACM Proceedings of the SIGCHI Conference on Human Factors in Computing Systems*, 2012, pp. 1911–1914.

[14] Wenjie Ruan, W., Q. Z. Sheng, and L. Shangguan, "AudioGest: Enabling Fine-Grained Hand Gesture Detection by Decoding Echo Signal," *UbiComp*, 2016.

[15] Pittman, C. R., and J. J. LaViola, "MultiWave: Complex Hand Gesture Recognition Using the Doppler Effect," *Proceedings of Graphics Interface 2017*, Edmonton, Alberta, May 16–19, 2017, pp. 97–106.

[16] Kalgaonkar, K., and B. Raj, "One-Handed Gesture Recognition Using Ultrasonic Doppler Sonar," *IEEE International Conference on Acoustics, Speech and Signal Processing*, 2009, pp. 1889–1892.

[17] Nandakumar, R., et al., "FingerIO: Using Active Sonar for Fine-Grained Finger Tracking," *Proceedings of the 2016 CHI Conference on Human Factors in Computing Systems*, San Jose, CA, May 7–12, 2016, pp. 1515–1525.

[18] Wang, Q., and Y. Liu, "Micro Hand Gesture Recognition System Using Ultrasonic Active Sensing Method," *IEEE SigPort*, 2016, http://sigport.org/1139.

[19] Li, G., et al., "Sparsity-Based Dynamic Hand Gesture Recognition Using Micro-Doppler Signatures," *Proceedings of IEEE 2017 Radar Conference*, 2017.

[20] Kim, Y. W., and B. Toomajian, "Hand Gesture Recognition Using Micro-Doppler Signatures with Convolutional Neural Network," *IEEE Access*, Vol. 4, 2016, pp. 7125–7130.

[21] Molchanov, P., et al., "Short-Range FMCW Monopulse Radar for Hand-Gesture Sensing," *Proceedings of IEEE Radar Conference*, 2015, pp. 1491–1496.

[22] Molchanov, P., et al., "Multi-Sensor System for Driver's Hand-Gesture Recognition," *Proceedings of the 2015 IEEE International Conference and Workshops on Automatic Face and Gesture Recognition*, Ljubljana, Slovenia, May 4–8, 2015.

[23] Wang, S., et al., "Interacting with Soli: Exploring Fine-Grained Dynamic Gesture Recognition in the Radio-Frequency Spectrum," *Symposium on User Interface Software and Technology*, 2016.

[24] Lien, J., et al., "Soli: Ubiquitous Gesture Sensing with Millimeter Wave Radar," *ACM Transactions on Graphics*, Vol. 35, No. 4, 2016, pp.142:1–142-19.

[25] https://atap.google.com/soli/.

[26] Zhang, S., et al., "Dynamic Hand Gesture Classification Based on Radar Micro-Doppler Signatures," *Proceedings of 2016 CIE International Conference on Radar*, Guangzhou, China, October 10–13, 2016.

[27] Ali, F., and M. Vossiek, "Detection of Weak Moving Targets Based on 2-D Range-Doppler FMCW Radar Fourier Processing," *Proceedings of 5th German Microwave Conference*, Berlin, Germany, 2010, pp. 214–217.

[28]   Sherman, S. M., and D. K. Barton, *Monopulse Principles and Techniques*, 2nd ed., Norwood, MA: Artech House, 2010.

[29]   Fishler, E., et al., "MIMO Radar: An Idea Whose Time Has Come," *Proceedings of the IEEE 2004 Radar Conference*, pp. 71–78.

[30]   Ram, S. S., and A. Majumdar, "High-Resolution Radar Imaging of Moving Humans Using Doppler Processing and Compressed Sensing," *IEEE Transactions on Aerospace and Electronic Systems*, Vol. 51, No. 2, 2015, pp. 1279–1287.

[31]   Ram, S. S., "Doppler Enhanced Frontal Radar Images of Multiple Human Activities," *Proceedings of the IEEE Radar Conference*, 2015.

[32]   Ram, S. S., S. Z. Gurbuz, and V. C. Chen, "Modeling and Simulation of Human Motion for Micro-Doppler Signatures," Chapter 3 in *Radar for Indoor Monitoring: Detection, Classification, and Assessment*, M. G. Amin, (ed.), Boca Raton, FL: CRC Press/Taylor & Francis Group, 2018, pp. 39–69.

7

Overview of the Micro-Doppler Radar System

Radar has been widely used for detecting and tracking targets with high accuracy at long range, day and night, and in all weather conditions. As a coherent radar, Doppler radar maintains high level of phase coherence for best tracking of Doppler information.

Micro-Doppler radar is a coherent Doppler radar that senses micro-motion-induced Doppler shifts and measures micro-Doppler signatures. A coherent radar transmits signals with all signal phases locked to a reference. Based on currently developed highly integrated radar systems, the narrowband continuous-wave (CW) radar and wideband frequency-modulated continuous-wave (FMCW) radar, which offer a great deal on the system compactness, flexibility, and low costs, are the best candidates for micro-Doppler radar.

7.1 Micro-Doppler Radar System Architecture

A simple architecture of a micro-Doppler radar system is shown in Figure 7.1. Since the radio frequency (RF) spectrum of the received signal is directly converted to baseband in the first downconversion, the radar receiver is a direct conversion or homodyne receiver. In homodyne receivers, the received

265

signal is often amplified to maintain a low noise figure and bandpass filtered for removing noise. Then the signal is directly converted to the baseband. The architecture is also called a zero intermediate frequency (IF) architecture [1, 2]. In a homodyne receiver, if it is not a quadrature mixer, then both sidebands of the signal are converted to the same baseband frequency region, which is known as the self-image effect. To avoid the self-image effect, a quadrature direct-conversion receiver is required. With the quadrature mixer, the received complex signal is mixed with a complex local oscillator such that only one sideband of the received complex signal is converted to the baseband frequency region without the self-image interference.

If the first downconverter is not a direct zero-IF conversion, it must have a second downconverter. Thus, the receiver is called a heterodyne receiver as shown in Figure 7.2 [3]. The received RF signal is first applied to a low-noise amplifier (LNA) and an image-reject filter. Then the image-rejected signal is mixed with the first oscillator (for example, a voltage-controlled oscillator (VCO)), called the VCO f_1, and results in an IF signal for further IF amplification and filtering. The IF-filtered signal is mixed with a second local oscillator (LO), f_{IF}, for producing baseband in-phase and quadrature phase (I and Q) outputs.

The basic issue in heterodyne receivers is the trade-off between the image rejection and interferer while selecting the IF [1]. As illustrated in Figure 7.3, in

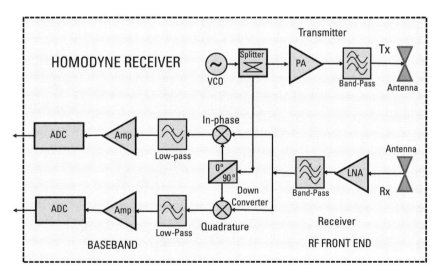

Figure 7.1 Architecture of a simple homodyne micro-Doppler radar system.

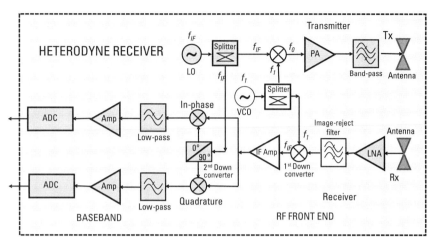

Figure 7.2 Architecture of a heterodyne radar system.

the heterodyne receiver, there is an undesired frequency band, which is equal to the RF minus twice the IF, called the image of the wanted frequency. Any interference and noise at the image frequency can interfere with reception of the desired signal frequency. The image-reject filter is for removing the interference within the image frequency region. The ability to reject interfering signals at the image frequency is measured by the image rejection ratio (IRR). This is the ratio of the output of a signal at the received frequency to the output of an equal-strength signal at the image frequency.

The homodyne receiver converts the band of interest directly to the zero-IF band and uses a lowpass filter to suppress interference. One of the advantages of the direct conversion over the heterodyne is that no image rejection filter is needed due to the zero-IF. However, the direct zero-IF conversion has number of troublesome issues that the heterodyne receiver does not have. These issues include the direct current (DC) offsets (i.e., LO leakage from the LO port to the mixer input and the LNA input to cause self-mixing, and interferer leakage from the LNA input to the LO port), I and Q imbalance, and flicker noise. Although the imbalance of I and Q has been a major issue compared with issues in the image-reject architecture, it is still less troublesome than issues in the image-reject filtering process [1].

In the homodyne receivers, the leakage problem is one of the troublesome issues. Because there is no separation in the time domain between the transmit and the receive signals, the continuous transmit and receive nature of the CW and FMCW signals results in the leakage from the transmitted

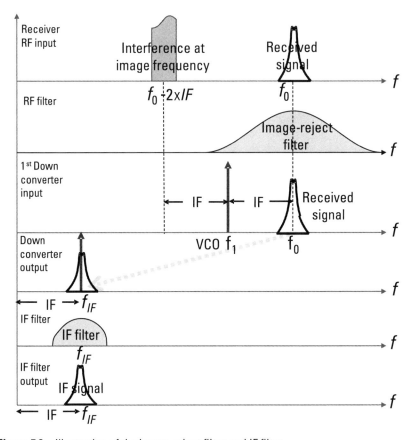

Figure 7.3   Illustration of the image-reject filter and IF filter.

signal directly to the receiver and thus reduces the performance of the radar receiver. To reduce the leakage, the use of physically separated antennas for transmit and receive is desirable. Sometimes a leakage cancellation circuit can also be used to suppress the leakage [4].

I and Q imbalance is also a common issue in the analog quadrature down-converter. An ideal downconverter should simply result in frequency shifting without introducing interferences. However, a real quadrature down-converter with I and Q imbalance not only downconverts the desired signal frequency, but also introduces its image interference and affect the demodulation performance. Especially in a wideband direct-converter, the frequency-dependent I and Q imbalance becomes a more severe problem [5–9].

7.2 Signal Waveforms for the Micro-Doppler Radar System

Radar signal waveforms must be suitable to the required function of the radar. Micro-Doppler radar is able to detect micromotion induced Doppler shifts and to measure micro-Doppler signatures. Thus, continuous waveform is the basic waveform for micro-Doppler radars. The CW radar transmits an electromagnetic wave with constant amplitude and frequency in an infinite time duration and measures only the Doppler frequency shift of moving objects without any range resolution.

The transmitted CW signal at a carrier frequency f_0 can be expressed by a complex form:

$$s_T(t) = A \cdot \exp\left\{ j2\pi f_0 t \right\} \tag{7.1}$$

The received signal becomes

$$s_R(t) = A_R \cdot \exp\left\{ j2\pi f_0(t - \tau) \right\} \tag{7.2}$$

where $\tau = 2R/c$ is the round-trip time delay.

If an object is moving with a radial velocity v_R, the received signal returned from the object becomes

$$s_R(t) = A_R \cdot \exp\left\{ j2\pi \left(f_0 + f_D \right)(t - \tau) \right\} \tag{7.3}$$

where $f_D = 2v_R/\lambda$ is the Doppler frequency shift.

The simplest CW radar system is a homodyne receiver system. In the homodyne receiver, the received signal $s_R(t)$ mixes directly with the transmitted signal $s_T(t)$, the Doppler shift frequency can be extracted from the mixed signal $s_M(t)$ through a lowpass filter (LPF) as depicted in Figure 7.4.

Instead of using a single mixer, by using an additional mixer with a 90° phase shifted version of the CW oscillator signal as shown in Figure 7.5, the sign of Doppler frequency shifts can be determined. The two mixers' scheme is called the quadrature demodulator with an in-phase (I) and a quadrature phase (Q) components. The output of the quadrature demodulator is a complex signal.

The I-component is the real part ($I = a_0 \cos\varphi$) and the Q-component ($Q = a_0 \sin\varphi$) is the imaginary part of a complex signal:

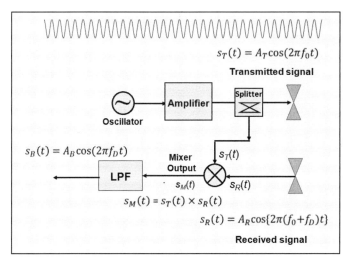

Figure 7.4   A block diagram of a homodyne CW Doppler radar front end.

$$S = I + jQ = a_0 \exp\{j\varphi\} \qquad (7.4)$$

where $a_0 = \sqrt{I^2 + Q^2}$ and $\varphi = \tan^{-1}(Q/I)$.

   A major limitation of the CW radar is that it cannot measure the range because it lacks a timing reference. In many cases, leveraging microrange and micro-Doppler signatures can capture more detailed features of micromotions. Thus, both high Doppler and high range resolutions may be required. To achieve a desired range resolution, the transmitted signal must have sufficient bandwidth. The wider bandwidth can be achieved through the modulation of the transmitted CW waveform, and the FMCW signal is a typical

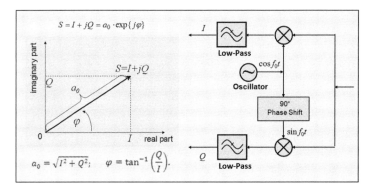

Figure 7.5   Quadrature demodulator.

modulation scheme. Other commonly used modulation schemes include the stepped-frequency waveform, pulse Doppler waveform, and coherent pulse (amplitude, phase, and frequency) modulation waveforms.

The FMCW radar transmits a continuous waveform modulated by a linear frequency modulation function. The frequency modulation provides necessary marks to allow both range and velocity information to be measured [10, 11]. The basic features of the FMCW radar include: (1) the time-delay range measurement is replaced by a beat frequency measurement through a fast Fourier transform (FFT); (2) the ability to measure very small distances comparable to the wavelength; (3) simultaneous measurement of range and relative velocity; (4) high processing gain with large time-bandwidth product such that quite low transmit powers may be used; (5) very high accuracy of range measurement; and (6) signal processing after mixing is performed at a low frequency band and thus simplifies the realization of the processing circuits.

A simple FMCW signal is the sawtooth modulation waveform as shown in Figure 7.6. The transmitted frequency $f_T(t)$ is a time-dependent function:

$$f_T(t) = f_0 + \frac{Bt}{T}, \quad (0 < t < T) \tag{7.5}$$

where f_0 is the initial frequency, B is the bandwidth of frequency sweep, and T is the sweep time duration. At the end of the sweep $t = T$, the frequency becomes $f_0 + B$.

The round-trip time delay τ of the signal returned from an object at a distance, R, is given by $\tau = 2R/c$, where c is the speed of the wave propagation. Thus, the received frequency becomes

$$f_R(t) = f_0 + B\frac{(t - \tau)}{T} \tag{7.6}$$

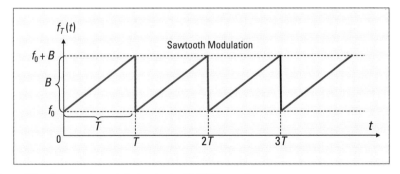

Figure 7.6 Sawtooth modulation in the FMCW waveform.

The mixer operates as the multiplication between its two input signals. By mixing the received signal with the transmitted signal, the sum and the difference of the frequency terms ($f_T + f_R$ and $f_T - f_R$) are generated at the output of the mixer. As illustrated in Figure 7.7, the difference term is a modulated low-frequency sinusoidal signal, called the beat frequency $f_B = |f_T - f_R|$, and the sum term is a high-frequency component $f_H = |f_T + f_R|$, where f_R is the receiving frequency and f_T is the transmitting frequency. The resulting signal is then processed by a lowpass filter to remove the sum term and keep the difference term, that is, the beat frequency f_B.

For a fixed range R, the beat frequency is independent of time:

$$f_B = f_T(t) - f_R(t) = \frac{B\tau}{T} = \frac{2BR}{cT} \qquad (7.7)$$

The process of generating a beat frequency from the return signal is illustrated in Figure 7.8.

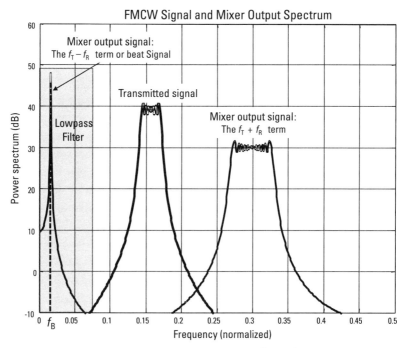

Figure 7.7 The output of the mixer: the beat frequency $|f_T - f_R|$, and a high-frequency component $|f_T + f_R|$ to be filtered by a lowpass filter.

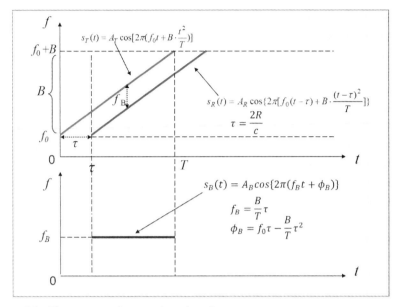

Figure 7.8   The process of generating a beat frequency.

Given the beat frequency f_B, the bandwidth B, and the sweep time T, the distance of the target can be calculated by

$$R = \left(\frac{c}{2B}\right) f_B T \tag{7.8}$$

To have an idea of the required sampling rate for the beat signal, given the bandwidth B, the sweep period T, and the range R, the beat frequency is $f_B = (B/T) \cdot (2R/c)$, where $2R/c$ is the time delay τ.

The range resolution ΔR is determined by $c/(2B)$ and the product of the sweep period and the beat frequency resolution, $\Delta f_B T$, that is,

$$\Delta R = \left(\frac{c}{2B}\right) \left(\Delta f_B T\right) \tag{7.9}$$

Because the beat frequency resolution Δf_B, is inversely proportional to the time interval of the Fourier transform, that is, $\Delta f_B = 1/T$. Thus, the range resolution in (7.9) is determined by the bandwidth $\Delta R = c/(2B)$.

In the case when the object is moving, a Doppler shift f_D is superimposed on the beat frequency. The transmitted and received frequency-time signals for the moving object are shown in Figure 7.9(a). Then the beat frequency is no longer a constant value. It is increased or decreased by the Doppler shift as shown in Figure 7.9(b). The beat frequency becomes: $f_B = f_R + f_D$, where f_R is the contribution of the object's range and f_D is the contribution of object's motion.

The maximum beat frequency is the one when both the object's range and the motion are maximal. Thus,

$$f_{B\max} = f_R + f_D = \frac{2B}{cT} R_{\max} + \frac{2}{\lambda} v_{\max} \tag{7.10}$$

For example, if the maximum range is 200m and the maximum velocity is 20 m/s, the maximum beat frequency becomes

$$f_{B\max} = \frac{2B}{cT} R_{\max} + \frac{2}{\lambda} v_{\max} = 667 \text{ kHz} \tag{7.11}$$

Thus, to satisfy the Nyquist criterion, the sampling rate for the analog-to-digital converter (ADC), f_S, must be greater than $2f_{B\max}$:

$$f_S \geq 2f_{B\max} = 1.334 \text{ MHz} \tag{7.12}$$

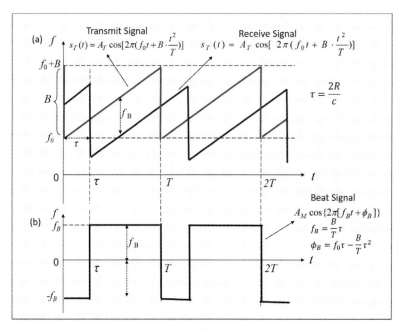

Figure 7.9 (a) The transmit and receive FMCW signals from a moving object, and (b) the beat signal.

Although the architecture of the homodyne receiver is simple, it also encounters problems, such as DC offsets and flicker noise. When the isolation between the signal input and the LO port of the mixer is not high enough, the LO leakage appearing at the mixer's signal input can cause self-mixing DC offset. Due to zero-IF, the flicker noise with $1/f$ spectral distribution in the devices can significantly corrupt the useful very low beat-frequency signals (such as heartbeat and respiration signals) near the DC component.

To overcome these problems in the homodyne architecture, the heterodyne receiver architecture may be used. The heterodyne receiver converts the received signal to an IF after the first downconversion. The IF signal is produced by mixing the received signal with the local VCO signal. Then the IF signal can be further amplified and filtered before performing target detection.

7.3   Resolution and Range Coverage

The range resolution, ΔR, indicates the ability of radar to distinguish two close targets. For FMCW radar, since the range, R, is deduced from the beat frequency, f_B, the range resolution, ΔR, can be derived from the beat frequency resolution, Δf_B. Because the length of each beat signal is less than the sweep time, T, the beat frequency resolution is usually defined as the 4-dB bandwidth of a monochromatic sinusoid with a rectangular window of length T:

$$\Delta f_B = \frac{1}{T} \qquad (7.13)$$

Thus, the range resolution, ΔR, can be derived consequently as

$$\Delta R = \frac{c}{2B} \qquad (7.14)$$

The radial velocity resolution, Δv_r, can be derived from the Doppler resolution Δf_D. Analogous to Δf_B, the Doppler resolution, Δf_D, is restricted by the window length in the time domain (i.e., the sweep time T); thus,

$$\Delta f_D = \frac{1}{T} \qquad (7.15)$$

Thus, the velocity resolution, Δv_r, can be deduced by

$$\Delta v_r = \left[\frac{c}{2f_0}\right]\Delta f_D = \left[\frac{c}{2f_0 T}\right] \qquad (7.16)$$

If a radar operating at $f_c = 24.125$ GHz continuously transmits FMCW signals with a bandwidth of $B = 250$ MHz within a sweep time $T = 1.0$ ms, the range resolution is $\Delta R = c/(2B) = 0.6$m, where c is the propagation speed. Assume that a group of sweeps, $n = 64$, is used to generate a range-Doppler map. Thus, the Doppler resolution becomes $\Delta f_D = 1/(nT) = 15.625$ Hz, or the velocity resolution is $\Delta v_r = [c/(2f_0)]\,\Delta f_D = 0.097$ m/s. The unambiguous velocity is $v_{max} = \pm c/(4Tf_0) = \pm 3.11$ m/s. If the sampling rate within a sweep is $f_S = 128$ kHz, the maximum range coverage is

$$R_{max} = \frac{c}{4B} f_s T = 38.4 \text{ m} \qquad (7.17)$$

7.4 Radar Range Equation

Figure 7.10 illustrates the power links of a two-way monostatic radar, where the transmit antenna and the receive antenna are colocated. The radar range equation is called the two-way tracking radar range equation or radar equation.

The signal-to-noise ratio (SNR) at the radar receiver is defined by the ratio of the detectable signal power to the noise power $SNR = P_R/P_N$, where the received signal power is

$$P_R = \frac{P_T G_T G_R \lambda^2 \sigma}{(4\pi)^3 R^4 L} \qquad (7.18)$$

and the receiver noise power or noise floor is

$$P_N = kT_0 B_n F_n \qquad (7.19)$$

where P_T is the peak transmit power of the transmitter, G_T is the transmit antenna gain, G_R is the receive antenna gain, λ is the wavelength, σ is the

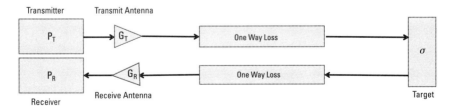

Figure 7.10 Power links of a two-way monostatic radar.

RCS of the target, R is the line-of-sight distance from the radar antenna to the target, L is the total loss, $k = 1.38 \times 10^{-23}$ Joule/K is the Boltzmann's constant, T_0 is the system temperature in Kelvin, B_n is the receiver noise bandwidth, and F_n is the receiver noise figure. Thus, the radar equation can be expressed by [12, 13]

$$SNR = \frac{P_R}{P_N} = \frac{P_T G_T G_R \lambda^2 \sigma}{(4\pi)^3 R^4 k T_0 B_n F_n L} \tag{7.20}$$

The important use of the radar range equation is to estimate the maximum detection range at which a target has a high probability of being detected by the radar. The criterion for detecting a target is that the SNR at the receiver is above a defined threshold. The radar range equation indicates that the SNR varies inversely with the fourth power of range. The detection range is defined as the range at which the SNR at the receiver can have a certain level.

The minimum detectable signal level S_{min} is determined by the noise power, P_N, in the receiver. In general, the minimum detectable signal level is set to the level of the noise power $S_{min} = P_N$. Thus, the maximum detection range is

$$R_{max} = \left\{ \frac{P_T G_T G_R \lambda^2 \sigma}{(4\pi)^3 S_{min} L} \right\}^{1/4} \tag{7.21}$$

where S_{min} is the receiver sensitivity or the minimum signal that can be detected:

$$S_{min} = \left(\frac{S}{N} \right)_{min} k T_0 B_n F_n \tag{7.22}$$

where $(S/N)_{min}$ is the required minimum signal-to-noise ratio for target detecting and processing.

7.4.1 CW Radar Range Equation

For a CW radar, the radar range equation must be modified based on (7.20). In the CW radar, the receiver output is passed to a bank of Doppler filters. If the output of a Doppler bin is higher than the detection threshold, the radar receiver declares a target is detected at this Doppler bin. Because the filter bank is implemented by taking an FFT, only data sets within a finite length of block can be processed at a time. The length of such block is normally referred to as the dwell interval T_{Dwell}, which is inversely proportional to the bandwidth of each individual bandpass filter:

$$B_n = \frac{1}{T_{\text{Dwell}}} \tag{7.23}$$

From the conventional radar equation, the SNR of the CW radar can be derived as

$$SNR = \frac{P_{\text{avg}} T_{\text{Dwell}} G_T G_R \lambda^2 \sigma}{(4\pi)^3 R^4 k T_0 F_n L} \tag{7.24}$$

where P_{avg} is the average transmitted power over the dwell interval.

Suppose that the dwell time interval T_{Dwell} is 0.256 second (−6 dBsec), the antenna gain G_T and G_R are both 20 dB, the transmit frequency is 5.8 GHz or the wavelength λ is 0.0517m (−12.9 dB), the propagation constant $\lambda^2/(4\pi)^3$ is −58.7 dB, the RCS of the target σ =1.0 sm (0 dBsm), the Boltzmann constant times the absolute temperature kT_0 is 4×10^{-21} w-s (−204 dBw-s), the noise figure F_n is 2.2 (3.4 dB), and the system loss L is 5 (7 dB), then the SNR become

$$SNR = 10\log_{10}\left(P_{\text{avg}}\right) + 155.9 - 40\log_{10}(R) \tag{7.25}$$

As shown in Figure 7.11, in a case where the average transmitted power P_{avg} is 15 dBm (31.6 mW) and the range R is 3,000m, the SNR will be 15 dB, which is above the required SNR of 13 dB for $P_d = 0.98$ and $P_{fa} = 10^{-4}$, according to the discussion of the required signal level in Section 7.4.3.

7.4.2 Receive Noise Floor

It is necessary to know the noise power to find the SNR. Noise comes from several sources: there is natural noise from the environment, thermal noise generated within the receiver itself, and man-made noise. The receive noise floor usually means the thermal noise floor. According to Boltzmann's law, any device with a temperature greater than absolute zero will generate thermal noise. The noise power per unit bandwidth is determined by kT_0, where $k = 1.38 \times 10^{-23}$ Joule/K is Boltzmann's constant and T_0 is the absolute temperature in Kelvin.

The noise floor is defined by the noise power at the receiver and given by $P_N = kT_0 B_n F_n$. For a room temperature $T_0 = 290$K, the noise power per hertz is

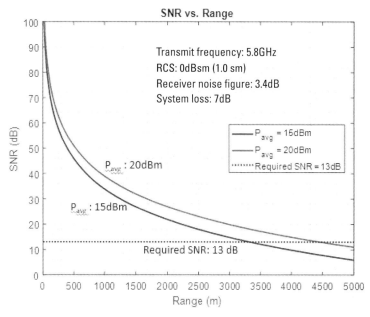

Figure 7.11 Received SNR versus target range for a 5.8-GHz CW radar.

$kT_0 = -228.7 + 24.6 = -204.1$ dBW $= -174.1$ dBm. If the noise figure F_n is 4 dB and the receiver noise bandwidth B_n is 1.0 MHz, the receiver noise floor is

$$P_N = kT_0 B_n F_n = -140 \text{ dBW} = -110 \text{ dBm}$$

7.4.3 Required Signal Level

Given a probability of detection P_d and probability of false alarm P_{fa}, the required SNR found in Figure 7.12. For $P_d = 0.98$ and $P_{fa} = 10^{-4}$, the required SNR is 13 dB above the noise level [12].

The required signal level P_R is estimated by adding the required SNR of 13 dB to the noise floor P_N, that is, $P_R = P_N + 13$ dB $= -110$ dBm + 13 dB $= -97$ dBm.

7.4.4 Received Signal Power

The received power is determined by the transmit power P_T, the antenna gains G_T and G_R, the wavelength λ, the radar cross-section (RCS) of the target σ, the system loss L, and the line-of-sight distance from the radar to the target R, given by (7.18):

$$P_R = \frac{P_T G_T G_R \lambda^2 \sigma}{(4\pi)^3 R^4 L}$$

For a C-band radar operating at 5.8 GHz and wavelength $\lambda = 0.0517$m, assuming the antenna gain $G_T = G_R = 20$ dB, the target's RCS is 1.0sm (0 dBsm), the loss L is 7 dB, and the propagation constant is $\lambda^2/(4\pi)^3 = -58.7$ dB, the received signal power is

$$P_R = \frac{P_T G_T G_R \lambda^2 \sigma}{(4\pi)^3 R^4 L} = 10\log_{10} P_T - 40\log_{10} R - 25.7 \qquad (7.26)$$

Thus, for a given transmit power P_T (17 dBm, 20 dBm, and 27 dBm), the received signal power is determined by the range R as shown in Figure 7.13. For example, if $P_T = 20$ dBm and $R = 100$m, the received signal power is -85.7 dBm, which satisfies the required signal level estimated in Section 7.4.3.

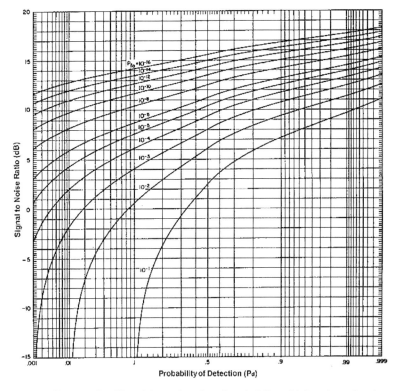

Figure 7.12 Given probability of detection P_d and probability of false alarm P_{fa}, the required SNR. (*After:* [12].)

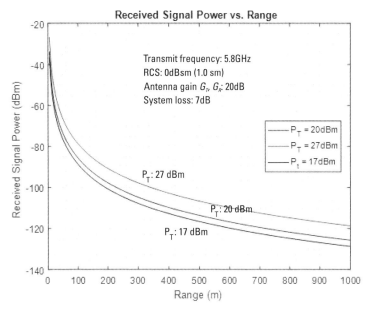

Figure 7.13 For a 5.8-GHz radar, antenna gain $G_T = G_R = 20$ dB, the target's RCS = 1.0sm, the loss $L = 7$ dB, given the transmit power P_T and target range, the calculated received signal power.

7.4.5 Receiver Sensitivity

The receiver sensitivity, S_{Rx}, is determined by the required minimum SNR, $(S/N)_{min}$ and the mean thermal noise power, which is defined by the receiver noise bandwidth B_n, receiver noise figure F_n, the absolute temperature T_0 at the receiver input, and the Boltzmann constant $k = 1.38 \times 10^{-23}$ Joule/K:

$$S_{Rx} = (S/N)_{min} k T_0 B_n F_n \qquad (7.27)$$

If the receiver is connected to an antenna, the sensitivity becomes the system operation sensitivity, S_{Oper}, which must consider the gain of the antenna system G:

$$S_{Oper} = \frac{(S/N)_{min} k T_0 B_n F_n}{G_R} \qquad (7.28)$$

For example, in a room temperature $T_0 = 290$K, $B_n = 1.0$ MHz, and receiver noise figure $F_n = 4$ dB, then $k T_0 B_n F_n = -110$ dBm. Therefore, with the minimum required $(S/N)_{min}$ of 13 dB and the antenna gain G_R of 20 dB, the receiver sensitivity is $S_{Rx} = -97$ dBm and the operation sensitivity is $S_{Oper} = -117$ dBm.

7.4.6   Receiver Dynamic Range

The receiver dynamic range is the ability to handle a range of signal strengths from the weakest detectable signal power, P_{min}, to the strongest signal power, P_{max}. The weakest detectable signal is determined by the receiver sensitivity. The strongest input signal power is determined by the saturation point of the receiver. The dynamic range is defined by the ratio of the input saturation point of power to the weakest detectable signal power: $DR = P_{max}/P_{min}$. The input saturation point of power can be determined by the third-order intercept point (IP3) because of the nonlinearity in the receiver. For receivers using analog-to-digital (A/D) converters, the input saturation point of power can be estimated from the number of bits used in the A/D converter (ADC) and from the input voltage range [11, 12].

The noise power in a receiver includes thermal noise, phase noise and quantization noise. The thermal noise power at the input of the ADC is $kT_0B_nF_nG_R$, where G_R is the gain of the receiver. Because the thermal noise at the receiver input is $kT_0B_nF_n = -110$ dBm and if the receiver gain is $G_R = 40$ dB, the thermal noise power at the input of the ADC should be -70 dBm. After range compression, the noise power is spread over all the available range cells. Thus, the SNR is increased by a factor equal to the time-bandwidth product of the radar signal, B_nT, where for FMCW waveform T is the sweep time duration. If $B_n = 1$ MHz and $T = 1.0$ ms, the time-bandwidth product is 1,000 or 30 dB. Thus, the thermal noise power in one resolution cell is

$$\left(P_{Thermal}\right)_{cell} = \frac{kT_0B_nF_nG_R}{B_nT} = \frac{kT_0F_nG_R}{T} = -100 \text{ dBm} \qquad (7.29)$$

Normally, a typical thermal noise power is about -90 dBm. The weakest detectable signal power, P_{min}, should be higher than the thermal noise power.

The quantization noise power has a significant contribution to the determination of the weakest detectable signal power. After A/D converting the demodulated analog signals to the digital domain, the maximum power and the minimum power are determined by the peak-to-peak input range of the A/D converter and the quantization noise of the digitizer.

For an A/D converter with 14-bit digitizers, the theoretical dynamic range can be calculated from the number of bits and the input voltage range.

If the peak-to-peak input range of the ADC is $2 \times V_{\text{max}_{\text{Peak}}} = 1.0\text{V}$, the maximum power level P_{max} that can be measured with a 50Ω input impedance is

$$P_{\text{max}} = 10\log\frac{\left(V_{\text{max}_{\text{RMS}}}\right)^2}{50} = 10\log\frac{\left(V_{\text{max}_{\text{Peak}}}/\sqrt{2}\right)^2}{50} = 10\log\frac{\left(0.5/\sqrt{2}\right)^2}{50} = 3.98 \text{ dBm}$$

(7.30)

Then the minimum power that can be measured due to the quantization noise of the digitizer is

$$P_{\text{min}} = 10\log\left[\frac{\left(V_{\text{Quant}_{\text{RMS}}}\right)^2}{50}\right]$$

(7.31)

In case of the A/D converter's intrinsic RMS noise $V_{\text{Quant}_{\text{RMS}}}$ is not available, it can be estimated by

$$V_{\text{Quant}_{\text{RMS}}} = \frac{2 \times V_{\text{max}_{\text{Peak}}}}{2^{14}\sqrt{2}} = 4.32 \times 10^{-5}$$

(7.32)

Then the minimum power is

$$P_{\text{min}} = 10\log\left[\frac{\left(V_{\text{Quant}_{\text{RMS}}}\right)^2}{50}\right] = -74.3 \text{ dBm}$$

(7.33)

Thus, for the A/D converter with 14-bit digitizer, the dynamic range for the quantization noise is

$$DR = P_{\text{max}} - P_{\text{min}} = 78.3 \text{ dB}$$

7.4.7 Maximum Detection Range

The maximum detection range is the maximum distance at which a target with a given RCS can be detected and processed. From the radar equation, the maximum detection range (7.21) is

$$R_{\text{max}} = \left[\frac{P_T G_T G_R \lambda^2 \sigma}{(4\pi)^3 k T_0 B_n F_n L(S/N)_{\text{min}}}\right]^{1/4}$$

To detect a human target with RCS of $\sigma = 1.0$ sm, a C-band radar is operating at 5.8 GHz with wavelength $\lambda = 0.0517$m. Based on Section 7.4.3, given $P_d = 0.98$ and $P_{fa} = 10^{-4}$, the required SNR is 13 dB or $(S/N)_{min} = 20$. If antenna gain $G_T = G_R = 20$ dB, the noise bandwidth B_n is 1.0 MHz, the noise figure F_n is 2.2, and the system loss L is 5, the maximum detection range is

$$R_{max} = \left[\frac{P_T G_T G_R \left(\lambda^2/(4\pi)^3 \right) \sigma}{kT_0 B_n F_n L (S/N)_{min}} \right]^{1/4} = \left[5.11 \times 10^{11} \times P_T \right]^{1/4}$$

where kT_0 is 4×10^{-21} W-s and the propagation constant $\lambda^2/(4\pi)^3$ is 1.35×10^{-6}.

Figure 7.14 gives the maximum detection range versus the transmit power for three different targets' RCS (0.01 sm for a bird, 1 sm for a human, and 10 sm for a car). When the transmitting power P_T is 0.1W, the maximum

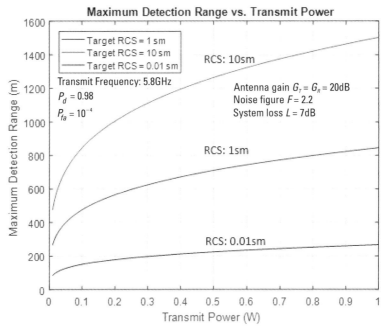

Figure 7.14   For 5.8-GHz radar and given $P_d = 0.98$ and $P_{fa} = 10^{-4}$, the antenna gain $G_T = G_R = 20$ dB, the noise bandwidth $B_n = 1.0$ MHz, the noise figure $F_n = 2.2$, and the system loss $L = 5$, the calculated maximum detection range versus the transmit power for a given RCS of a target.

detection range is 198m for a human (1 sm), 351m for a car (10 sm), and 52.5m for a bird (0.01 sm).

7.5 Data Acquisition and Signal Processing

7.5.1 Noise Sources

The noise comes from thermal noise, phase noise and quantization noise. As explained in Section 7.4.6, the thermal noise power at the input of the ADC is about −70 dBm. After range compression, the noise power is spread over all the available range cells and the SNR is increased by a factor equal to the time-bandwidth product of the radar signal. The thermal noise power in one resolution cell is about −100 dBm. The major contribution of the phase noise in the receiver is due to the coupling of the transmitted signal at the mixer. The mixer's isolation between two inputs determines the phase noise power. Only a fraction of the transmitted signal is used to mix with the received signal. By reducing the oscillator phase noise and increasing the mixer isolation, the phase noise in the receiver can be improved. The typical required oscillator phase noise at a frequency offset 1 MHz relative to the carrier frequency is about −110 dBc/Hz. The required mixer isolation is about −40 dB. The typical phase noise power in one resolution cell $(P_{\text{Phase}})_{\text{cell}}$ is about −175 dBW. As discussed in Section 7.4.6, the quantization noise of the digitizer is −81 dBm or −111 dBW.

Flicker noise always occurs in the receiver. Since it features with a $1/f$ power spectral density, flicker noise can be significant at very low frequencies.

7.5.2 Digital Data Acquisition

Digital data acquisition is the process of sampling received signals and converting them into digital bits for further signal conditioning and processing as shown in Figure 7.15.

The I and Q signal outputs from the mixer is sampled by two A/D converters. The critical parameters of the A/D converter are its resolution bits and sampling frequency f_s. During A/D conversion, the continuous analog signal is quantified and the number of quantified levels determines the resolution bits. For example, with the resolution of 14 bits, the quantization levels are $2^{14} = 16,384$. However, the resolution of quantization also determines the quantization noise that may determine the minimum SNR of the receiver. According

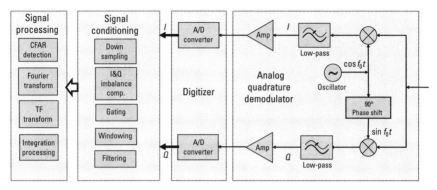

Figure 7.15   Digital data acquisition digitizes the analog signals for further signal conditioning and processing.

to the sampling theory, the sampling frequency f_s must be at least twice higher than the bandwidth of baseband signal for avoiding frequency aliasing.

7.5.3   Signal Conditioning

Signal conditioning is manipulation of signals for the next stage of processing. It includes I-Q imbalance correction, downsampling, filtering, thresholding, range-gating, frequency-windowing, and any other processes required to make the output suitable for further processing.

For the CW signal, the sampling rate in the digitizer is usually high enough for covering the maximal Doppler frequency shift. Therefore, in most cases, a downsampling of the original digitized signal is needed. The downsampling rate depends on the required maximal micro-Doppler frequency. Figure 7.16 shows a digitized CW signal waveform and the downsampled signal. The digitized CW signal is the output of the receive signal mixed with the transmit CW signal and then lowpass-filtered and digitized. Figure 7.16(a) is the original digitized baseband signal and Figure 7.16(b) shows the downsampled baseband signal at a suitable decimate factor.

After downsampling, the baseband signal is ready for the next stage of processing. Depending on applications, windowing, filtering, thresholding, and time-frequency analysis may be applied.

In some cases where range resolution is required, instead of using narrowband CW signals, a wideband signal waveform must be used. The FMCW signal is a popular one often used in the micro-Doppler radar. The FMCW signal usually has sawtooth-modulation or triangular-modulation. The sawtooth-modulation is used in most micro-Doppler applications.

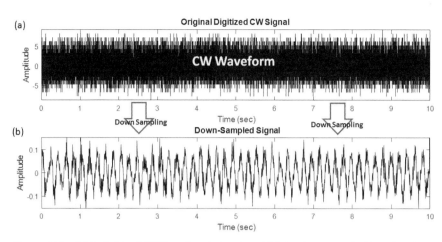

Figure 7.16 (a) A digitized CW baseband signal, and (b) the downsampled baseband signal.

For an FMCW radar, if data is collected during N times the sweep time duration T, that is, $N \times T$, and there are M samples during each sweep time T, then the Doppler resolution is determined by the total number of sweeps N and the range resolution is determined by the number of time samples M during each sweep. Because both the Doppler and the range information are needed, the radar collected raw data must be rearranged into a two-dimensional (2-D) $N \times M$ matrix, as shown in Figure 7.17. After conversion to a 2-D array, the data is ready for the next stage of processing, such as windowing, filtering, and time-frequency analysis.

Compared with the CW radar, for a given data collection time duration, the FMCW radar gained range resolution, but lost Doppler resolution by M times.

7.5.4 In-Phase and Quadrature Imbalance and Its Compensation

I and Q imbalance is commonly seen in analog quadrature downconverter. Imbalance introduces imaginary components of the spectrum in the output of the converter such that limits the demodulation performance. The imbalance in either the gain or phase of the I and Q-component signals create a shift from the normal I-Q graph. Since the I and Q imbalance is frequency-dependent, it makes the compensation of the imbalance more difficult in wideband direct-conversion receivers.

Various publications exist regarding the compensation of the I and Q imbalance in communication systems [4–9]. Assuming a single-frequency RF

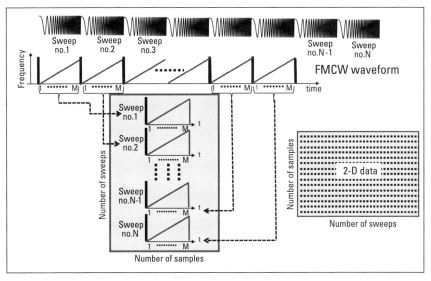

Figure 7.17   A digitized FMCW signal rearranged as 2-D data.

signal to be converted to a baseband signal, the effect of amplitude and phase imbalance discussed in [9] can be described in the form of

$$
\begin{cases}
I(t) = \cos(2\pi f t) \\
Q(t) = \sin(2\pi f t) \\
I'(t) = (1 + A)\cos(2\pi f t) \\
Q'(t) = \sin(2\pi f t + \psi)
\end{cases}
\tag{7.34}
$$

where $I(t)$ and $Q(t)$ are the outputs of an ideal downconverter and f is the frequency of the baseband signal. The imbalanced I and Q components of the quadrature downconverter are $I'(t)$ and $Q'(t)$, where A is the amplitude imbalance and ψ is the phase imbalance. To compensate the imbalance, the proposed procedure in [9] is to introduce two correction operations: (1) operation P to rotate the $Q'(t)$; and (2) operation E to scale the $I'(t)$. The compensated components $I''(t)$ and $Q''(t)$ are related to the quadrature downconverter output $I'(t)$ and $Q'(t)$ by

$$
\begin{bmatrix} I''(t) \\ Q''(t) \end{bmatrix} = \begin{bmatrix} E & 0 \\ -P & 1 \end{bmatrix} \begin{bmatrix} I'(t) \\ Q'(t) \end{bmatrix}
\tag{7.35}
$$

where $E = \cos(\psi)/(1 + A)$ and $P = \sin(\psi)/(1 + A)$. Thus, the final compensated components $I''(t)$ and $Q''(t)$ are

$$\begin{cases} I''(t) = \cos(\psi)\cos(2\pi ft) \\ Q''(t) = \cos(\psi)\sin(2\pi ft) \end{cases} \tag{7.36}$$

However, this method must have the value of A and ψ before making imbalance compensation. Thus, a test signal must be used to find out the A and ψ.

Another similar algorithm for I and Q imbalance compensation was proposed in [14], where the amplitude imbalance A and the phase imbalance ψ can be estimated from the imbalanced $I'(t)$ and $Q'(t)$ themselves.

Assume the outputs of an ideal downconverter are $I(t) = \cos(2\pi ft)$ and $Q(t) = \sin(2\pi ft)$, where f is the baseband frequency of the signal. Thus, in a realistic direct conversion receiver, the outputs of the downconverter are modeled by

$$I'(t) = A\cos(2\pi ft) \tag{7.37}$$

and

$$Q'(t) = \sin(2\pi ft + \psi) \tag{7.38}$$

where A in the I-channel is the amplitude imbalance and ψ in the Q-channel is the phase imbalance. Thus, the idea $I(t)$ and $Q(t)$ can be expressed by the imbalanced $I'(t)$ and $Q'(t)$ by

$$\begin{bmatrix} I(t) \\ Q(t) \end{bmatrix} = \begin{bmatrix} 1/A & 0 \\ -\sin(\Psi)/[A\cos(\psi)\} & 1/\cos(\psi) \end{bmatrix} \begin{bmatrix} I'(t) \\ Q'(t) \end{bmatrix} \tag{7.39}$$

where A, $\sin(\psi)$ and $\cos(\psi)$ can be computed by

$$A = \left[2\langle I'(t)I'(t)\rangle\right]^{1/2} \tag{7.40}$$

$$\sin(\psi) = \frac{2}{A}\langle I'(t)Q'(t)\rangle \tag{7.41}$$

and

$$\cos(\psi) = \left[1 - \sin^2(\psi)\right]^{1/2} \tag{7.42}$$

Based on the above analysis, imbalance compensation can be conducted simply by digital signal processing with the I and Q correction matrix in (7.39).

However, the above algorithms are based on a single-frequency signal and thus may fail for a wideband signal. In practice, some empirical algorithms may be used to partially improve imbalance compensation for wideband signals as provided MATLAB source code in this book [15]. Figure 7.18 shows collected human walking data by an X-band 9.8-GHz micro-Doppler radar. A person walks away from and then returns back to the radar during

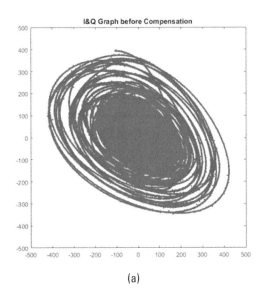

(a)

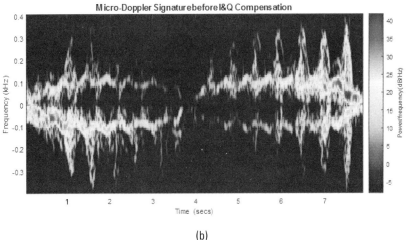

(b)

Figure 7.18 (a, b) Before and (c, d) after applying the empirical I and Q imbalance compensation algorithm.

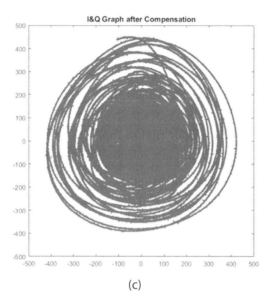

(c)

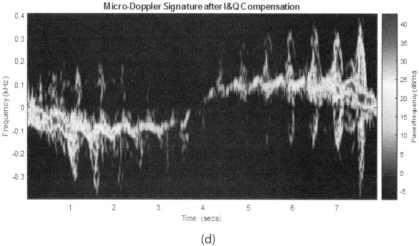

(d)

Figure 7.18 *(Continued)*

8 seconds of the time duration. Because of the I and Q imbalance, Figure 7.18(a) plots the I and Q imbalanced graph and Figure 7.18(b) is the I and Q imbalanced micro-Doppler signature. Figures 7.18(c, d) show, after applying the empirical imbalance compensation algorithm, the I and Q graph and the micro-Doppler signature, respectively. It shows that the empirical algorithm works fine in this example, but it is not always like this.

References

[1]   Razavi, B., "Design Considerations for Direct-Conversion Receivers," *IEEE Transactions on Circuits and Systems—II: Analog and Digital Signal Processing*, Vol. 44, No. 6, 1997, pp. 428–435.

[2]   Svitek, R., and S. Raman, "DC Offsets in Direct-Conversion Receivers: Characterization and Implications," *IEEE Microwave Magazine*, September 2005, pp. 76–86.

[3]   Gruz, P., H. Gomes, and N. Carvalho, "Receiver Front-End Architectures: Analysis and Evaluation," in *Advanced Microwave and Millimeter Wave Technologies: Semiconductor Devices Circuits and Systems*, M. Mukherjee, (ed.), London, U.K: InTech, 2010.

[4]   Kok, P. W., "Evaluation of Wideband Leakage Cancellation Circuit for Improved Transmit-Receive Isolation," Thesis, The Naval Postgraduate School, Monterey, CA, December 2011.

[5]   De Witt, J. J., "Modeling, Estimation and Compensation of Imbalances in Quadrature Transceivers," Ph.D. Dissertation, Stellenbosch University, South Africa, 2011.

[6]   Churchill, F. E., G. W. Ogar, and B. J. Thompson, "The Correction of I and Q Errors in a Coherent Processor," *IEEE Transactions on Aerospace and Electronic Systems*, Vol. AES-17, No. 1, 1981, pp. 131–137.

[7]   Green, R. A., R. Anderson-Sprecher, and J. W. Pierre, "Quadrature Receiver Mismatch Calibration," *IEEE Transactions on Signal Processing*, Vol. 47, No. 11, 1999, pp. 3130–3133.

[8]   Stormyrbakken, C., "Automatic Compensation for Inaccuracies in Quadrature Mixers," Master of Science in Electronic Engineering Thesis, University of Stellenbosch, December 2005.

[9]   Easton, D., J. Snowdon, and D. Spencer, *Quadrature Phase Error in Receivers*, Project Report, Dept. of Electronics and Computer Science, University of Southampton, December 2008.

[10]   Rahman, S., and D. A. Robertson, "Coherent 24 GHz FMCW Radar System for Micro-Doppler Studies," *Proceedings of SPIE*, Vol. 10633, Radar Sensor Technology XXII, 1063301, 2018.

[11]   Wang, L., *60GHz FMCW Radar System with High Distance and Doppler Resolution and Accuracy*, Graduation Paper, Dept. of Electrical Engineering, Eindhoven University of Technology, The Netherlands, 2010.

[12]   Skolnik, M.I., *Radar Handbook*, 3rd ed., New York: McGraw-Hill, 2008.

[13]   Skolnik, M.I., *Introduction to Radar Systems*, 3rd ed., New York: McGraw-Hill, 2001.

[14]   Ellingson, S.W., *Correcting I-Q Imbalance in Direct Conversion Receiver*, ElectroScience Laboratory, The Ohio State University, 2003.

[15]   Research/Micro-Doppler Analysis and I&Q Imbalance Correction, http//www.ancortek.com.

8

Analysis and Interpretation of Micro-Doppler Signatures

Human and animal body and limb motions contain a rich source of information on body movements, actions, and intentions. The ability of visual systems to perceive object motion from sparse input is the phenomenon called biological motion perception. Visual system can easily retrieve information and identify a body from its motion pattern. Why the human visual system can easily identify an object through its motion and how exactly the biologically and psychologically relevant information is encoded in the motion pattern are not clear yet.

Numerous studies on biological motion perception have been ongoing for a long period of time. Experimental studies have indicated how observers can recognize human figures through their body and limb movement and even identify the type of a motion pattern (such as walking or running). Certainly, these tremendous experimental studies can help for the selection of meaningful features from radar micro-Doppler signatures and thus to identify types of motion patterns and finally recognize a person through his or her motion pattern.

The micro-Doppler signature is a pattern of time-varying Doppler frequency shifts where the frequency shift and its rate of change directly relate to the radial velocity and acceleration of the motion. Thus, methods for extracting motion kinematic features from micro-Doppler signatures are extremely important. If the kinematic features of an object can be obtained, the study of human motion perception can be directly applied to reconstruct the body movement of the object.

In this chapter, some useful results on visual perception of biological motion and identification through visual motion patterns are briefly introduced. Based on the knowledge of biological motion perception, methods of how to decompose a micro-Doppler signature into components associated with structural body parts, how to interpret a micro-Doppler signature, and how to select features from micro-Doppler components are discussed.

8.1 Biological Motion Perception

The pioneering experimental work on biological motion perception by Johansson in 1973 and 1976 proved that visual perception of point-light displays (PLD) attached to a limited number of major joints of a walking person can help to immediately recognize the structure of the human body [1, 2]. Experiments verified that a 0.2-second interval is sufficient for an observer to recognize the figure of a walking person and a 0.4-second interval is sufficient to identify the type of a motion pattern.

Following Johansson's work, researchers investigated how the structural body movement and corresponding kinematic information help for classifying the type of a movement, recognizing the person, and even identifying the gender of the person [3–8].

It is well known that people can recognize a friend based on familiarity cues, such as face, hairstyle, or even gait. Facial recognition is a most common method for recognizing a person. However, whether and how a person can be recognized by his or her motion pattern are questionable. Researchers have been investigated this topic for years [4, 9, 10]. Cutting and Kozlowski conducted the first experiment on recognizing persons from their biological motions through the PLDs [4]. They recorded the gait patterns from number of persons who were familiar to each other, but not familiar with their figures of PLDs. In this experiment, these persons repeatedly watched the PLDs from individual persons. Finally, the PLD experiment showed that not only the PLDs are sufficient for indicating the presence of a human structure as Johansson's demonstration [1, 2], but also contain sufficient information for identifying individuals through their motion patterns. This experiment demonstrated that, from learning of PLDs, people can differentiate previously unknown persons through the way they walk. However, this experiment was still unclear whether there were specific factors or parameters that are intrinsic to the walkers for the identification.

Researchers have also studied that, from the visual motion perception point of view, which parts of structural information and kinematic information are relevant for identifying the types of the movement and the identity of a person. Troje et al. have made significant contributions to biological motion perception and motion decomposition [7, 8, 11]. They showed that the perception of biological motion depends on the links between joints (i.e., the point-light pairs), the trajectories of the individual PLDs (called the local motion information), and the information on the entire figure of the PLDs across a larger spatiotemporal interval (called the global motion information). It was found that the recognition of biological motion is not solely based on the local motion information, but also based on the perception of the global figure. Figure 8.1 illustrates a running person represented by a limited number of point displays.

Troje also proposed methods based on principal component analysis [7] and Fourier analysis [8] to decompose human walking data into structural and kinematic information. In their methods, an averaged posture represented by local motions of the PLDs along with the period of gait was used as the human walking pattern. After subtracting the posture from the kinematic data, the residual kinematic data was decomposed by the PCA algorithm. It was found that the first four principal components are the major components representing the walker's posture. The observers are more dependent on the kinematic

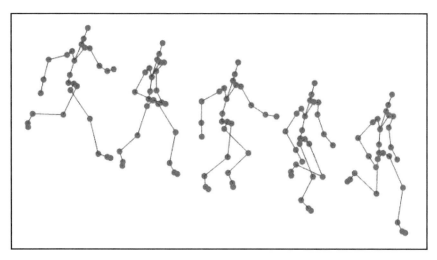

Figure 8.1 A running person represented by a limited number of point displays.

information for the recognition of individual walkers. They also applied the Fourier decomposition method, which can be used to separately manipulate various attributes (such as size, shape, and gait frequency) and to examine the differential influence of these parameters on the identification of individuals.

Researchers found that the visual kinematics of the PLDs carry information on actions [2, 3, 12, 13], emotions [14–18], and even the gender of the walker [6–8, 19–21]. From any movement of the PLDs (such as knocking, lifting, and waving), human observers can discriminate between the passive affect and the aggressive affect. Experiments also indicated that the ability of distinguishing between different emotions from the PLDs was based on the information of motion kinematics (such as velocity, acceleration, and jerk).

After knowing biological motion perception from the PLDs, the next step is how to extract the necessary kinematic and structural features from micro-Doppler signatures. As is well known, the micro-Doppler signature is a characteristic of the intricate frequency modulations generated from each component part of an object and represented in the joint time and Doppler frequency domain. Based on the experimental result on human motion perception from the PLDs of human body components, it can be inferred that motion kinematics of human body components in micro-Doppler signatures can also carry information on human actions and emotions.

As is also well known, the micro-Doppler signature of a human body movement is a superposition of micro-Doppler components from individual body parts. Thus, it is necessary to decompose a micro-Doppler signature into components such that each of them corresponds to the motion from each body component part. The decomposition of human micro-Doppler signatures therefore becomes a challenging problem that may lead to the identification of human actions and emotions through micro-Doppler signatures.

8.2   Decomposition of Biological Motion

To decompose motion data into structural and kinematic information components, the Fourier-based transform and principal component analysis (PCA) are the first ones to be considered.

The Fourier transform transfers the motion data into spectral components in the frequency domain, where the center frequency component (i.e., the Doppler shift) encodes the structural information of the moving body's geometry and the dynamic sideband frequency components (i.e., the micro-Doppler shifts) encode the kinematic motion information.

In the PCA method, after subtracting the averaged posture, the residual kinematic data are used for the PCA decomposition. The PCA is a common tool in statistical data analysis [22], which calculates the eigenvalue decomposition of the data covariance matrix. The PCA can reduce the dimensionality of the data and reveal the internal structure of the data through the data variance. Thus, for biological motion data, the PCA method actually results in a discrete Fourier decomposition, which is optimal in the sense of maximizing variance with a minimum number of components [8].

However, the decomposition of biological motion is not directly decomposing the micro-Doppler function. It is a challenge to decompose a micro-Doppler signature of a biological motion into those components and each of them is associated with an individual body part.

8.2.1 Statistics-Based Decomposition

The statistically independent decomposition includes PCA, singular value decomposition (SVD), and independent component analysis (ICA) [22–25].

PCA is a tool in statistical data analysis and uses eigenvectors with the largest eigenvalues to obtain a set of function bases so that the original function can be represented by a linear combination of these bases. The function bases found by the PCA are uncorrelated (i.e., they cannot be linearly predicted from each other). However, higher-order dependencies still exist in the PCA, and these bases are not optimally separated.

SVD is a generalization of the PCA and can decompose a nonsquared matrix, which is possible to directly decompose the time-frequency distribution without using a covariance matrix.

ICA is important to the decomposition of signals and has many practical applications. A well-known example of the application of ICA is the "cocktail party problem," in which the ICA can extract an underlying speech separated from people talking simultaneously in a room. ICA was originally used for separating mixed signals into independent components, called blind source separation (BSS). ICA minimizes the statistical dependence between basis feature vectors and searches for a linear transformation to express a set of features as a linear combination of statistically independent function bases. As is well known, independent events must be uncorrelated, but uncorrelated events may not be independent. PCA only requires the components to be uncorrelated. However, ICA is independent and accounts for higher-order statistics. Thus, it is a more powerful feature representation than PCA. PCA, just like the Fourier analysis, is basically a global component analysis, whereas ICA, like the time-frequency analysis, is basically a localized component analysis.

Certainly, the statistically decomposed components are not these mono-components that correspond to micro-Doppler signatures generated from individual body parts. The desired decomposition method for micro-Doppler signatures should be the one that is based on the physical structural components of human body parts.

8.2.2 Decomposition of Micro-Doppler Signatures in the Joint Time-Frequency Domain

Micro-Doppler signatures are usually represented in the joint time and frequency domain. They can be a monocomponent function or a multicomponent function. The multicomponent function is a superposition of multiple monocomponent functions. Each monocomponent signature corresponds to a single component body part of an object.

There are many methods by selecting a basis (kernel) function to decompose a joint time-frequency distribution into independently controllable components. These kernel functions are just localized elementary basis functions in the joint time and frequency domain, but not associated with any individual component body part of the object.

Another type of decomposition is the statistically independent decomposition such as PCA or SVD and ICA. These decomposed components are uncorrelated or independent, but not necessarily associated with individual component body parts of an object. These monocomponent signatures are often correlated or dependent due to synchronized locomotion of the individual component body parts, such as a human walking.

EMD, introduced in Chapter 1, is a signal decomposition method that decomposes an original signal into component waveforms, called the intrinsic mode functions (IMF), that are modulated in their amplitude and frequency by searching all of the oscillatory modes in the signal [26]. One property of the IMFs is that different IMFs do not have the same instantaneous frequency at the same time instant.

EMD has been applied to micro-Doppler signatures for extracting radar signal components generated by rotating or vibrating body structures [27]. Although the EMD can extract informative harmonic components from human motion data, there is still no connection to any structural human body part.

Therefore, the most useful but difficult decomposition method for micro-Doppler signatures should be the one that is based on the physical structural components of human body parts.

8.2.3 Physical Component-Based Decomposition

The micro-Doppler signature of a moving body does not look like the visual perception of a body motion in a three-dimensional (3-D) space. However, it is directly linked to the kinematic information of the moving body.

For better performing target classification, recognition and identification based on its micro-Doppler signatures, the first step is to decompose the micro-Doppler signature into multiple monocomponent signatures associated with physical parts of the target. For example, for a micro-Doppler signature of a human gait, the decomposed monocomponent signatures should be associated with the torso, arms, legs, and feet. Thus, the classifier may be able to identify human actions, emotions, and even the gender information. Similarly, for a micro-Doppler signature of a hand gesture, the decomposed monocomponent signatures should be associated with the palm, thumb, index, middle, and other fingers.

A human observer can easily track a monocomponent signature from a multicomponent micro-Doppler signature. However, it is not easy for a machine to track and isolate any individual monocomponent signature from a multicomponent micro-Doppler signature. Thus, an effective algorithm on the physical component-based decomposition is still an open issue.

8.2.3.1 Parametric and Nonparametric Decomposing Micro-Doppler Signatures

In [28, 29], a framework was proposed for decomposing micro-Doppler signatures into components that are associated with the physical parts of a human body. A micro-Doppler signature of a walking human and the decomposed signature, called motion curves, that correspond to the different physical parts of the human body are shown in Figure 8.2. The decomposition algorithm used is described as follows. First, the class of the motion must be predetermined from the given micro-Doppler signature by examining the structure of variations of the upper and lower envelopes in the micro-Doppler signature being analyzed. This was accomplished by means of a Markov chain-based inference algorithm [28]. The advantage of focusing on the upper and lower envelopes is that their micro-Doppler structure remains relatively invariant to the distortions inherent in the acquisition process. Inferring the motion class yields valuable prior knowledge about the internal time-frequency structure of the signal to be analyzed; such knowledge is obtained by prior simulation studies in [30]. Conditioned on this knowledge, the motion class of the remaining body parts is determined by the following process.

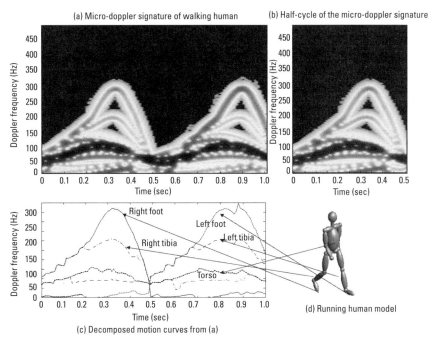

(a) Micro-doppler signature of walking human   (b) Half-cycle of the micro-doppler signature

(c) Decomposed motion curves from (a)

(d) Running human model

Figure 8.2   (a) Micro-Doppler signature of a walking person, (b) half-cycle of the micro-Doppler signature of the walking person, (c) the decomposed motion curves from the micro-Doppler signature in (a), and (d) the model of the walking person.

The micro-Doppler signature being analyzed is segmented into consecutive different half-cycles such that each half-cycle corresponds to roughly half the period of motion, as shown in Figure 8.2(b). This segmentation is accomplished by means of a nonlinear optimization program that is solved via dynamic programming. Thus, the motion curves extracted for each half-cycle can then be concatenated to form the overall motion sequence of the human. To extract the motion curves for each half-cycle, the significant local maxima correspond to the maximal time cell of the half-cycle. The maximal time cell is at the time coordinate where the upper/lower envelope attains global maxima. These significant local maxima are determined by means of the same nonlinear optimization algorithm described in [28]. Given this, the local corresponding maxima at the remaining time cells are determined by means of a partial tracking algorithm described in [28, 29]. The concatenation of the corresponding local maxima at different time cells yields the motion curves for each of the body parts. The number of local maxima to be computed (i.e., the number of body parts for which to determining the motion curves) is known by prior knowledge obtained by simulation studies and associated with the inferred motion class. Then the initial set of motion curve estimates obtained are refined by means of a Gaussian g-Snakes-based quality measure as described in [28, 29], where Gaussian g-Snakes provide a model of the micro-Doppler structure of the human gait motion that enables a blind assessment of the quality of the motion curves estimates in Figure 8.2(c) [28, 29].

Figure 8.3(a) shows the micro-Doppler signature of a running person. Figure 8.3(b) shows the corresponding decomposed motion curves extracted for this running motion.

The framework demonstrated the signature decomposition only for human walking and running. The extracted motion curves are also very limited. Thus, more generalized algorithms may be expected that can work with more complex micro-Doppler signatures and decompose into more motion curves. To achieve this goal, it is necessary to understand the nonlinear interactions between the motion dynamics and the motion curves, including the efficient representation of complex motion dynamics and their effect on the micro-Doppler signature.

8.2.3.2 Viterbi Algorithm for Decomposing Micro-Doppler Signatures

A key issue in decomposing a multicomponent micro-Doppler signature into multiple monocomponent signatures is to estimate the true instantaneous frequencies (IFs) and track the time-varying IF trajectory for finding each monocomponent signatures.

As a form of dynamic programming, the Viterbi algorithm can be used to estimate the hidden state in a joint time-frequency domain and has achieved some level of success [30–34]. Based on the Viterbi algorithm [35], a method of IF estimation was proposed for reducing the error caused by strong noise and decomposing a micro-Doppler signature in a joint time-frequency domain through detecting IFs at each time and finding the best paths to the next IFs.

To extract micro-Doppler signals in the joint time-frequency domain, a time-varying frequency filter instead of traditional time-invariant filter was used in [32]. To determine a time-frequency masking function, the Viterbi algorithm is used to estimate the IFs. After multiplying the time-frequency transform of a micro-Doppler signal with the time-frequency mask filter, the micro-Doppler signal can be reconstructed by applying an inverse time-frequency transform.

The Viterbi algorithm is an effective tool for estimating IFs in the time-frequency domain. A simple IF estimation based on detecting the time-frequency maxima is expressed by

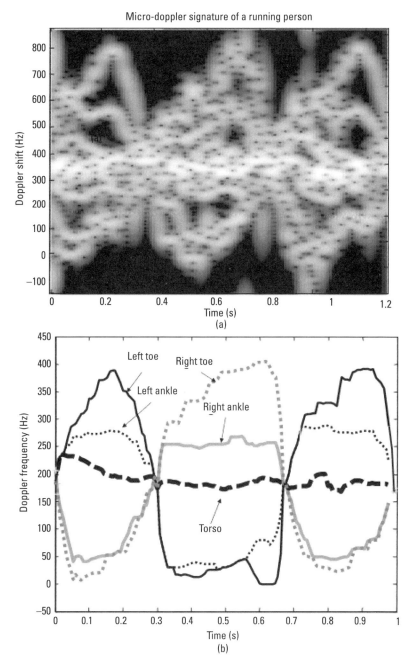

Figure 8.3   (a) Micro-Doppler signature of a running person, and (b) the decomposed motion curves from the micro-Doppler signature of the running person in (a).

$$IF(n) = \arg\max_{k} TF(n,k)$$

where $TF(n, k)$ is the time-frequency matrix, n and k are the time index and the frequency index, respectively. By using this method, all IF points can be independently detected at corresponding time slices. However, this is not suitable in cases of a strong noise environment where the estimated IF can deviate from its true IF. Thus, an algorithm based on the Viterbi algorithm and time-frequency distribution may be applied. The Viterbi algorithm for estimating the IFs is based on following assumptions: (1) the time-frequency values at the IFs located points are large enough, and (2) the IF variation between two consecutive points is not extremely large. Thus, the IF estimator should minimize the corresponding sum of the path penalty functions [30–33].

8.3 Extraction of Features from Micro-Doppler Signatures

Because the most often performed human locomotion is walking, extracting motion features from micro-Doppler signatures of a walking human is a basic method for analyzing human motion.

In this section, the global human walking model [36] is used to perform a simulation of human walking to generate micro-Doppler signatures of a walking human. Because of the simulation study, each human body part can be isolated from others and the corresponding monocomponent micro-Doppler signature can be observed. Thus, the corresponding structural and kinematic information of each human body part may be extracted.

Based on the global human walking model, a radar operating at Ku-band at 15 GHz is located at $(X_1 = 10m, Y_1 = 0m, Z_1 = 2m)$. A person at 1.8-m height is walking from a point at $(X_0 = 0m, Y_0 = 0m, Z_0 = 0m)$ toward the radar with a radial velocity of $v_R = 1.0$ m/s, as illustrated in Figure 8.4.

As mentioned in Chapter 4, because the global human walking model is built based on averaging parameters from experimental measurements, it is just an averaging human walking model without incorporating information about personalized motion features.

Figure 8.5 shows the micro-Doppler signature of the walking person. The micro-Doppler components of the feet, tibias, clavicles, and torso are marked in the figure. The average of the torso Doppler frequency shift is about 133 Hz. Figure 8.6 shows the corresponding micro-Doppler components of the feet, tibias, radius, and torso, respectively.

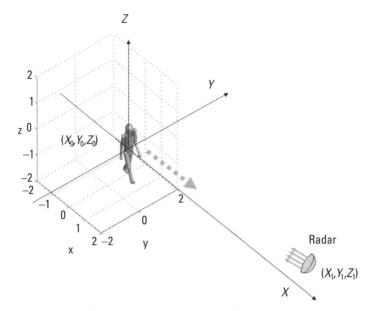

Figure 8.4   Geometry of a walking person and the radar.

8.4   Estimation of Kinematic Parameters from Micro-Doppler Signatures

Micro-Doppler signatures of human body movement show strong reflections from the torso due to its larger radar cross-section (RCS). From the micro-Doppler signature of the torso shown in Figure 8.7, the average velocity, the

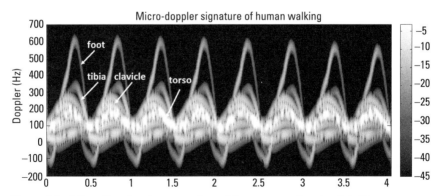

Figure 8.5   Micro-Doppler signature of a walking person using the global human walking model.

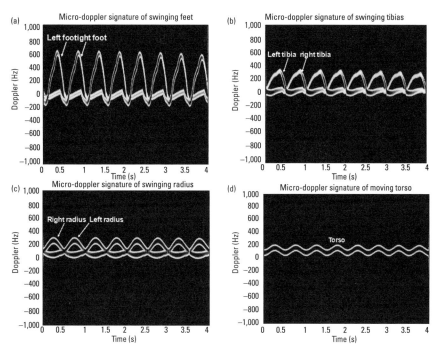

Figure 8.6   Micro-Doppler components of: (a) feet, (b) tibias, (c) radius, and (d) the torso.

oscillation cycle and the amplitude of the torso oscillation can be measured. From the simulation using the global human walking model for a Ku-band radar operating at 15 GHz, the torso oscillating velocity is from 1.0 m/s and going up to 1.7 m/s, as shown in Figure 8.7. The mean value of the torso velocity is $v_{torso} = f_D \lambda/2 = 133 \times 0.02/2 = 1.33$ m/s. The oscillation period of the torso is 0.5 second or, in other words, the frequency of the torso oscillation is 2 Hz.

The corresponding lower leg (tibia) motion parameters are shown in Figure 8.8. The average tibia velocity is $v_{tibia} = 129 \times 0.02/2 = 1.29$ m/s, the cycle of the tibia oscillation is 1 Hz, and the maximum amplitude of the tibia oscillation velocity is about 3.2 m/s.

The corresponding foot motion parameters are shown in Figure 8.9. The average foot velocity is $v_{foot} = 127 \times 0.02/2 = 1.27$ m/s, the cycle of the foot oscillation is 1 Hz, and the maximum amplitude of the foot oscillation velocity is about 5.9 m/s. Half the foot oscillation cycle is the foot forward motion, and the other half of the cycle is the foot in contact with the ground, which makes the average velocity lower. The foot motion has the highest velocity, which

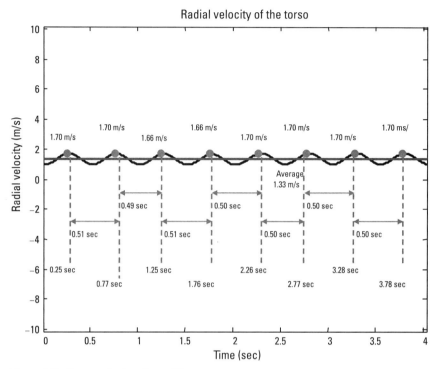

Figure 8.7 Torso velocity of a walking person.

is approximately four to five times the average foot velocity. The oscillation frequency of the torso is two times the tibia or the foot oscillation frequency because the torso accelerates while either foot swings.

8.5   Identifying Human Body Movements

After decomposing and feature extraction, the final procedure is to identify the human body movement. Before getting into this, it is necessary to clarify terminologies of the classification, recognition, and identification [37, 38]. Although people often use the three terminologies ambiguously, for radar applications, the target classification often requires different levels of categories.

The first level is to identify the nature or the class of the detected target, such as a human or an animal, a ground vehicle or an aircraft. The second level is to identify the type, such as whether the ground vehicle is a trunk or a school bus. The third level is to confirm the model or a specific individual in

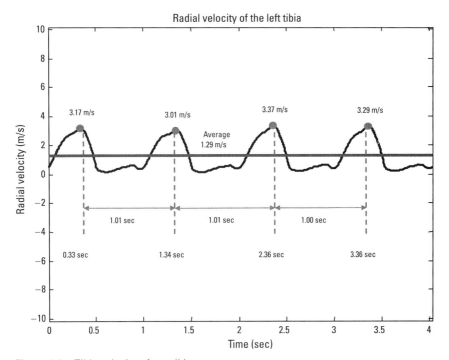

Figure 8.8 Tibia velocity of a walking person.

detail from others, such as whether the tank is a T-72 tank or an M-60 tank. Therefore, the first level is classification, the second level is called recognition, and the third level is called identification.

The purpose of identifying a human body movement is to determine the model of the movement or identify an individual movement different from others, such as walking, running, or jumping, after classifying the movement as being made by humans. Existing approaches to identify human body movements in the computer vision are divided into two different types: the exploitation of the structural information, and the use of the motion information [39, 40]. The motion-based algorithm uses the moments of human body parts, eigenvectors, and hidden Markov models (HMMs) [41, 42]. Although these motion-based algorithms have been successful in many scenarios, due to the lack of structural information, in some other scenarios they could be worse than the methods that mainly use structural information [43].

The pattern of a human body movement is a characteristic of the observed human body movement. Different types of movements generate different types of motion patterns. The micro-Doppler signature is a type of motion pattern

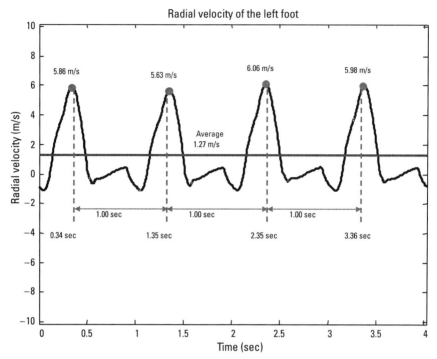

Figure 8.9   Foot velocity of a walking person.

[44]. The structural characteristic of human body movements can be extracted from the motion pattern. It can be a distinctive measurement, a transformation, or a structural component. The features extracted from motion patterns are the key to the identification of a human body movement.

8.5.1   Features Used for Identifying Human Body Movements

It has been demonstrated that human body movement could be identified through the point-light displays [1]. The experiment showed that by placing point lights on the joints of human body segments and filming human body movements in a dark room, the point-light displays of the human body movement can provide a vivid impression of human body motion. This means that the hierarchical structure of body segments and the constraints on the segments and joints are the characteristic features of human body movements. These characteristic features are the key features for identifying human body movements.

To completely identify a human body movement, the first step is to determine the type of the human body motion, such as walking, running, jumping, or crawling; the second step is to identify the phase of the motion, such as stance phase or swing phase when walking; and the third step is to predict possible motion continuation. With available decomposed human body motion components, such as a component that corresponds to the motion of an individual human body part, it is possible to completely identify the movement and the motion phase and even to predict possible intention.

8.5.2 Anomalous Human Behavior

Anomaly detection is a technique to identify aberrant behavior and unusual occurrences that do not conform to normal behavior. The key to detecting an anomaly is knowing how a normal behavior functions and what the normal behavior looks like. The anomaly detection technique can be applied to wide areas from health care to cybersecurity and from safety to military surveillance.

However, the detection of anomalous behavior is a difficult task. Simple statistical tools, such as the mean, quantiles, mode, and distribution, may work for identifying irregularities in some simple cases, but it may not work well in many other cases. Machine learning-based approaches, such as the k-nearest neighbor algorithm and the support vector machine (SVM), are popular techniques for the anomaly detection.

Although the visual observation of human actions may help to understand human mental activities, it is difficult to know how a human action is decoded into intentions, what properties of a human action make it special, and how these properties are organized to make cognitive representations.

Anomaly detection provides an alarm signal for a strange behavior. To perform the detection, an activity profile of normal behaviors is used to compare with current activity. Anything that deviates from the normal behavior is classified as anomalous. However, how to interpret human activities, track these movements, select appropriate features, and detect events of interest are still open issues.

As well known, any human activity is accomplished by a sequence of complicated motion of a human body part and described by a rotation about an axis through the center of mass and a translation of the center of mass. Thus, the kinematic information on a human body part is the sum of kinematics of rotation and the translation of the body part. Thus, the kinematic information is a discriminative feature that may have higher discriminatory power for identifying human activities. The kinematic information can be extracted from the decomposed monocomponent micro-Doppler signatures of individual human body parts.

To detect a human anomalous behavior, it must model the human behavior. HMMs are popular in computer vision as an activity recognition algorithm [40–42]. They have been successfully used to recognize hand gestures and facial expressions and to classify activities in visual surveillance systems. An approach used in the automated visual surveillance is to classify normal activities using a set of discrete HMMs, each trained to recognize one activity, and label the unrecognized activities as unusual [41].

The Markov model is a probabilistic technique for learning and matching activity patterns. Each type of activity for human events may be characterized by a family of event trajectories. Each family can be represented as an HMM in which states represent regions, the prior probabilities measure the likelihood of an event starting in a particular region of the process, and the transitional probabilities capture the likelihood of progression from one state to another across the whole process.

8.6 Summary

In this chapter, based on the biological motion perception, methods of decomposing micro-Doppler signatures into monocomponent signatures of body parts have been discussed. The physical component-based decomposition method opened a window for showing the feasibility of extracting kinematic features from micro-Doppler signatures. More effective algorithms for the signature decomposition are still needed.

In the computer vision, using motion kinematic features to predict possible human actions and identify anomalous human behaviors is a difficult task. If kinematic parameters of human body parts can be extracted from micro-Doppler signatures, just like in computer graphics, those animated human models using sensor motion captured data, the animated human body parts may be generated from the captured micro-Doppler signatures. From here, it is possible to identify human actions and even anomalous behaviors from micro-Doppler signatures.

References

[1] Johansson, G., "Visual Perception of Biological Motion and a Model for Its Analysis," *Perception & Psychophysics*, Vol. 14, 1973, pp. 201–211.

[2]   Johansson, G., "Spatio-Temporal Differentiation and Integration in Visual Motion Perception," *Psychological Research*, Vol. 38, 1976, pp. 379–393.

[3]   Dittrich, W. H., "Action Categories and the Perception of Biological Motion," *Perception*, Vol. 22, 1993, pp. 15–22.

[4]   Cutting, J. E., and L. T. Kozlowski, "Recognizing Friends by Their Walk: Gait Perception Without Familiarity Cues," *Bulletin of the Psychonomic Society*, Vol. 9, 1977, pp. 353–356.

[5]   Loula, F., et al., "Recognizing People from Their Movement," *Journal of Experimental Psychology: Human Perception and Performance*, Vol. 31, 2005, pp. 210–220.

[6]   Barclay, C. D., J. E. Cutting, and L. T. Kozlowski, "Temporal and Spatial Factors in Gait Perception That Influence Gender Recognition," *Perception and Psychophysics*, Vol. 23, 1978, pp. 145–152.

[7]   Troje, N. F., "Decomposing Biological Motion: A Framework for Analysis and Synthesis of Human Gait Patterns," *Journal of Vision*, Vol. 2, 2002, pp. 371–387.

[8]   Troje, N. F., "The Little Difference: Fourier Based Synthesis of Gender-Specific Biological Motion," in *Dynamic Perception*, R. Weurtz and M. Lappe, (eds.), Berlin: AKA Verlag, 2002, pp. 115–120.

[9]   Beardsworth, T., and T. Buckner, "The Ability to Recognize Oneself from a Video Recording of One's Movements Without Seeing One's Body," *Bulletin of the Psychonomic Society*, Vol.18, 1981, pp. 19–22.

[10]   Stevenage, S. V., M. S. Nixon, and K. Vince, "Visual Analysis of Gait as a Cue to Identity," *Applied Cognitive Psychology*, Vol. 13, 1999, pp. 513–526.

[11]   Chang, D. H. F., and N. F. Troje, "Characterizing Global and Local Mechanisms in Biological Motion Perception," *Journal of Vision*, Vol. 9, No. 5, 2009, pp. 1–10.

[12]   Pollick, F. E., C. Fidopiastis, and V. Braden, "Recognising the Style of Spatially Exaggerated Tennis Serves," *Perception*, Vol. 30, 2001, pp. 323–338.

[13]   Sparrow, W. A., and C. Sherman, "Visual Expertise in the Perception of Action," *Exercise and Sport Sciences Reviews*, Vol. 29, 2001, pp. 124–128.

[14]   Atkinson, A. P., et al., "Emotion Perception from Dynamic and Static Body Expressions in Point-Light and Full-Light Displays," *Perception*, Vol. 33, 2004, pp. 717–746.

[15]   Dittrich, W. H., et al., "Perception of Emotion from Dynamic Point-Light Displays Represented in Dance," *Perception*, Vol. 25, 1996, pp. 727–738.

[16]   Heberlein, A. S., et al., "Cortical Regions for Judgments of Emotions and Personality Traits from Point-Light Walkers," *Journal of Cognitive Neuroscience*, Vol. 16, 2004, pp. 1143–1158.

[17]   Pollick, F. E., et al., "Estimating the Efficiency of Recognizing Gender and Affect from Biological Motion," *Vision Research*, Vol. 42, 2002, pp. 2345–2355.

[18]   Pollick, F. E., et al., "Perceiving Affect from Arm Movement," *Cognition*, Vol. 82, 2001, pp. B51–61.

[19]   Kozlowski, L. T., and J. E. Cutting, "Recognizing the Sex of a Walker from a Dynamic Point-Light Display," *Perception & Psychophysics*, Vol. 21, 1977, pp. 575–580.

[20]   Mather, G., and L. Murdoch, "Gender Discrimination in Biological Motion Displays Based on Dynamic Cues," *Proceedings of the Royal Society of London Series B*, Vol. 258, 1994, pp. 273–279.

[21]   Sparrow, W. A., et al., "Visual Perception of Human Activity and Gender in Biological-Motion Displays by Individuals with Mental Retardation," *American Journal of Mental Retardation*, Vol. 104, 1999, pp. 215–226.

[22]   Jolliffe, I. T., *Principal Component Analysis*, Springer Series in Statistics, New York: Springer-Verlag, 1986.

[23]   Golub, G. H., and C. Reinsch, "Singular Value Decomposition and Least Squares Solutions," *Numerische Mathematik*, Vol. 14, No. 5, 1970, pp. 403–420.

[24]   Hyvarinen, A., and E. Oja, "Independent Component Analysis: Algorithms and Applications," *Neural Networks*, Vol. 13, No. 4-5, 2000, pp. 411–430.

[25]   Chen, V. C., "Spatial and Temporal Independent Component Analysis of Micro-Doppler Features," *IEEE 2005 International Radar Conference*, Washington, D.C., May 2005.

[26]   Huang, N. E., et al., "The Empirical Mode Decomposition and the Hilbert Spectrum for Nonlinear and Non-Stationary Time Series Analysis," *Proc. Roy. Soc. Lond. A*, Vol. 454, 1998, pp. 903–995.

[27]   Cai, C. J., et al., "Radar Micro-Doppler Signature Analysis with HHT," *IEEE Transactions on Aerospace and Electronic Systems*, Vol. 46, No. 2, 2010, pp. 929–938

[28]   Raj, R. G., V. C. Chen, and R. Lipps, "Analysis of Radar Human Gait Signatures," *IET Signal Processing*, Vol. 4, No. 3, 2010, pp. 234–244.

[29]   Raj, R. G., V. C. Chen, and R. Lipps, "Analysis of Human Radar Dismount Signatures Via Parametric and Non-Parametric Methods," *IEEE 2009 Radar Conference*, Pasadena, CA, May 2009.

[30]   Stankovic, L., et al., "Signal Decomposition of Micro-Doppler Signatures," Chapter 10 in *Radar Micro-Doppler Signature—processing and applications*, V. C. Chen, D. Tahmoush, and W. J. Miceli, (eds.), Radar Seriers 34, London, U.K.: IET, 2014, pp. 273–328.

[31]   Djurović, I., V. Popović-Bugarin, and M. Simeunović, "The STFT-Based Estimator of Micro-Doppler Parameters," *IEEE Transactions on Aerospace and Electronic Systems*, Vol. 53, No. 3, 2017, pp. 1273–1283.

[32]   Li, P., D. C. Wang, and L. Wang, "Separation of Micro-Doppler Signals Based on Time Frequency Filter and Viterbi Algorithm," *Signal, Image and Video Processing*, Vol. 7, No. 3, 2013, pp. 593–605.

[33]   Mazurek, P., "Estimation of Micro-Doppler Signals Using Viterbi Track-Before-Detect Algorithm," *2017 22nd International Conference on Methods and Models in Automation and Robotics*, 2017, pp. 898–902.

[34] Abdulatif, S., et al., "Real-Time Capable Micro-Doppler Signature Decomposition of Walking Human Limbs," *2017 IEEE Radar Conference*, 2017, pp. 1093–1098.

[35] Forney, G. D.,"The Viterbi Algorithm," *Proc. IEEE*, Vol. 61, No. 3, 1973, pp. 268–278.

[36] Boulic, R., N. Magnenat-Thalmann, and D. Thalmann, "A Global Human Walking Model with Real-Time Kinematic Personification," *The Visual Computer*, Vol. 6, No. 6, 1990, pp. 344–358.

[37] Anderson, S. J., "Target Classification, Recognition and Identification with HF Radar," *RTO SET Symposium Proceedings on Target Identification and Recognition Using RF System*, MP-SET-080, Oslo, Norway, October 11–13, 2004.

[38] Chen, V. C., "Evaluation of Bayes, ICA, PCA and SVM Methods for Classification," *RTO SET Symposium Proceedings on Target Identification and Recognition Using RF System*, MP-SET-080, Oslo, Norway, October 11–13, 2004.

[39] Gavrila, D., "The Visual Analysis of Human Movement, A Survey," *Computer Vision and Image Understanding*, Vol. 73, No. 1, 1999, pp. 82–98.

[40] Aggarwal, J., and Q. Cai, "Human Motion Analysis: A Review," *Computer Vision and Image Understanding*, Vol. 73, No. 3, 1999, pp. 428–440.

[41] Sunderesan, A., A. Chowdhury, and R. Chellappa, "A Hidden Markov Model Based Framework for Recognition of Humans from Gait Sequences," *Proc. of 2003 IEEE Intl. Conf. on Image Processing*, Vol. 2, 2003, pp. 93–96.

[42] Lee, L., and W. Grimson, "Gait Analysis for Recognition and Classification," *Proc. Intl. Conf. on Automatic Face and Gesture Recognition*, Vol. 1, 2002, pp. 155–162.

[43] Veeraraghavan, A., A. R. Chowdhury, and R. Chellappa, "Role of Shape and Kinematics in Human Movement Analysis," *Proc. of 2004 IEEE Conference on Computer Vision and Pattern Recognition (CVPR)*, Vol. 1, 2004, pp. 730–737.

[44] Li, W., B. Xiong, and G. Kuang, "Target Classification and Recognition Based on Micro-Doppler Radar Signatures," *2017 Progress in Electromagnetics Research Symposium*, Singapore, November 19–22, 2017.

9

Summary, Challenges, and Perspectives

The primary purpose of this book is to introduce the principle of the micro-Doppler effect in radar and to provide a simple and easy tool for simulation of micro-Doppler signatures of radar signals reflected from objects with micro motions. Based on examples provided in this book, readers may modify and extend these examples to various applications of their interest.

Because of recent development in applications of micro-Doppler effect in radar [1–4], there are many emerging applications and new advances available. In this second edition of the book, three chapters (Chapters 5, 6, and 7) are added to introduce applications of the micro-Doppler effect to vital sign monitoring and hand gesture recognition and to provide an overview of the requirement and architectures of micro-Doppler radar systems.

9.1 Summary

The micro-Doppler effect in radar can be observed by any coherent Doppler radar; and the micro-Doppler signatures can be generated by a joint time-frequency analysis. As discussed in Chapter 1, the instantaneous frequency analysis method is a simple way to generate time-varying micro-Doppler signatures. However, it is only appropriate for analyzing monocomponent signals.

315

In cases of multicomponent signals, the instantaneous frequency analysis is not suitable and, thus, a joint time-frequency distribution method must be used.

Doppler radar is to measure the radial velocity of a moving object. If the object has no radial velocity, Doppler radar is unable to measure the velocity of the object. While an object moves along a curved path, when its radial velocity decreases, its angular velocity must increase. Thus, for completely describing the object's motion, its angular velocity must be determined. A technique for measuring the angular velocity is the correlation interferometry used in radio astronomy [5]. It was found that in a similar way to the relationship between the Doppler frequency shift and the radial velocity, the interferometric frequency shift is proportional to the angular velocity, the carrier frequency, and the baseline of the interferometer.

To simulate micro-Doppler signatures in radar, a suitable radar cross-section (RCS) model for description of objects and an accurate kinematic model for description of micromotions are needed. In Chapters 3 and 4, commonly used methods for modeling objects and for description of motion equations are given. Micro-Doppler signatures of typical rigid body and nonrigid body motions are demonstrated. MATLAB source codes for modeling rigid and nonrigid body motions and algorithms for calculations of micro-Doppler signatures are provided in the book. The POFACET is a simple RCS prediction software package written in MATLAB codes and is for calculation of monostatic and bistatic RCSs, which can be downloaded [6, 7].

Because of recent progress on human vital sign detection and hand gesture recognition using Doppler radars, applications of micro-Doppler signatures to vital sign detection and to hand gesture recognition are introduced and in Chapters 5 and 6.

Since the detection of vital signs is based on the detection of micromotions on the human chest wall, a high sensitivity of Doppler radar to surface vibrations has brought great chance for radar detection of vital signs.

Micro-Doppler signatures can also be applied to hand gesture recognition. When hands and fingers are making different gestures, their corresponding micro-Doppler signatures are different in their shapes, intensities, and durations. Through micro-Doppler signature analysis, feature extraction, pattern recognition, and machine learning, hand gestures can be recognized [8–11].

Recently developed highly integrated RF chips make it possible to build low cost, low-power and compact Doppler radar systems for sensing of micro motions. An overview of the requirement and system architecture for building a micro-Doppler radar system is given in Chapter 7.

In Chapter 8, some useful results on biological motion perception are introduced, which may help radar engineers to consider how to use those results in radar micro-Doppler signature analysis and in identification of targets of interest.

9.2 Challenges

Micro-Doppler signatures can be used to extract the kinematic features and to identify an object of interest. However, how to effectively extract and correctly interpret the features and how to associate them with structural components of the object is still a challenge. Although a human observer can easily track a monocomponent signature from a multicomponent micro-Doppler signature, it is not an easy task for a machine to track and isolate any individual monocomponent signature from a multicomponent micro-Doppler signature. An effective decomposition algorithm for decomposing a complex micro-Doppler signatures into physical component-based monocomponent signatures is still an open issue. The success in solving meaningful features of these structural components may lead to significant improvement in classification, recognition and identification in general, and may even identify the intention and behavior from a biological motion.

9.2.1 Decomposing Micro-Doppler Signatures

Micro-Doppler signature is usually a superposition of multiple monocomponent signatures. Each of them is associated with an individual structural component of an object. In current decomposition methods, some of them decompose a time-frequency distribution into independently controllable localized components, such as the matching pursuit method and the adaptive Gabor method [12, 13]. However, those decomposition methods do not take into account the mechanism that how a complex micro-Doppler signature is formed from signatures of individual physical components of an object. The empirical model decomposition (EMD) method does decompose micro-Doppler signatures into monocomponents, but those monocomponents have no connection to any physical component of the object. A useful decomposition method should be the one that associates the decomposed components with the physical components of the object.

The decomposition of micro-Doppler signature is a challenge. A recent method to track a monocomponent signature in a multicomponent micro-Doppler signature is based on the Viterbi algorithm and time-frequency distribution [14, 15]. Finding an effective algorithm for decomposing multicomponent micro-Doppler signatures into monocomponents is still an open issue.

9.2.2   Feature Extraction and Kinematic Parameter Estimation from Micro-Doppler Signatures

The possible features and kinematic parameters that can be extracted from micro-Doppler signatures include the time information, the frequency or period of micromotion, the magnitude and the sign of Doppler frequency shifts, the position and moving direction, the linear velocity and acceleration, and the angular velocity and acceleration. In a monostatic radar system, the measured Doppler frequency is propotional to the radial line-of-sight (LOS) velocity component of the moving object being observed. In a bistatic radar system, the measured Doppler frequency is related to the velocity projected onto the line of the bistatic bisector. Therefore, for accurately locating the position and measuring the true moving direction and the true velocity, two or more monostatic radars or an interferometric radar are needed.

Figure 9.1 depicts a simple case for measuring the true velocity of a moving target using two monostatic radars. To describe the geometry of the radars and the target in two-dimensional (2-D) Cartesian coordinates, a moving target is initially at the location at (X_0, Y_0), and two radars are located at (X_1, Y_1) and (X_2, Y_2), respectively. Thus, the angle α between the LOS of radar no. 1 and the LOS of the radar no. 2 is known.

For simplicity, it is assumed that the moving target is moving toward the inside of the scalene triangle. The radial velocities measured by radar no. 1 and radar no. 2 are V_{R1} and V_{R2}, respectively. The altitude line 1 in the triangle is the line that is perpendicular to the radial velocity V_{R1}; the altitude line 2 is the line that is perpendicular to the radial velocity V_{R2}. Then the cross-point of the altitude line 1 and the altitude line 2 determines the true velocity V of the moving target.

Based on the geometry of the scalene triangle and the positions of the target, radar no. 1, and radar no. 2, by the measurement of radial velocities V_{R1} and V_{R2}, it is easy to calculate the true velocity magnitude and the angle of arrival.

Thus, if the initial position of any physical part of a target can be measured and its monocomponent micro-Doppler signature of the target can be extracted, then the basic kinematic parameters (positions and true velocities)

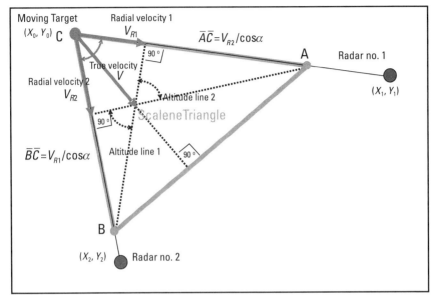

Figure 9.1   Estimation of the true velocity V of a moving target based on two radial velocities V_{R1} and V_{R2}.

of each physical part of the target can be estimated from the two monostatic radars.

To completely describe a target (such as a human) motion in a three-dimensional (3-D) Cartesian coordinate system, linear kinematic parameters (linear position, linear velocity, and linear acceleration) are the most important motion kinematic parameters. These parameters define the manner in which the human body part changes over time. Linear velocity describes the rate of the position change with respect to time, and linear acceleration describes the rate of velocity change with time. These three kinematic parameters will be used to determine the characteristics of a target movement. The other three kinematic parameters are the angular kinematics, which include the angular position (orientation) of a target body part, angular velocity, and angular acceleration of the body part. Because a target may consist of a number of parts, the measurement of joint angles between two parts are useful for describing target movement. These three angular kinematic parameters can be used jointly with three linear kinematic parameters to completely describe motions of physical parts of the target. By carefully handling rotation and translation, the 3-D trajectories of these joints between body parts can be obtained. These linear and angular kinematic parameters of the target movement are used for classification, recognition, and identification of the target of interest.

9.3   Perspectives

Since the history of the research on micro-Doppler effect in radar is relatively short, many aspects of research topics are still open and remain to be exploited. These topics include multistatic micro-Doppler analysis, micro-Doppler signature-based classification, deep learning using micro-Doppler signatures, through-the-wall micro-Doppler analysis, aural methods in micro-Doppler-based discrimination, and micro-Doppler signatures for detecting target in sea clutter.

9.3.1   Multistatic Micro-Doppler Analysis

Multistatic radar has multiple transmitter/receiver nodes over distributed locations [16]. Each node in the multistatic system can have only one transmitter or receiver. Multistatic radar is considered a combination of multiple bistatic radars that observes targets from different aspects. Thus, the informative data acquired from targets is increased because of the multiple aspects viewing. Multistatic radar takes the advantage of spatial diversity such that different aspects of a target can be viewed simultaneously and more complete Doppler and micro-Doppler shifts can be observed. Multistatic radar fuses the received data in the receivers, and the performance of the fusion depends on the degree of spatial coherency between channels, the topology of the system, the number of targets and their spatial locations, and the complexity of the targets.

Multistatic micro-Doppler signature depends on the topology of the system and the location and motion of the target [17]. The information encoded in the micro-Doppler signature may not be linearly increased with the number of channels used in the system because of the possible correlations between channels. If the cross-correlation between channels is one with no time delay, this indicates that the received signals in the two channels are the same and, thus, the second channel contains no additional information. If the two channels are not correlated, the second channel contains different information and is an informative channel. The value of Doppler shifts detected in each channel depends on the topology of the system, the target location, and its moving direction. By fusing information captured from multiple channels, the location, moving direction, and velocity of the target can be measured.

With the increased information, the radar performance on target recognition is expected to be improved.

9.3.2 Micro-Doppler Signature-Based Classification, Recognition, and Identification

Target classification, recognition, and identification are extremely important research topics in radar. Commonly used statistical algorithms include linear discriminant, naïve Bayes classifier, support vector machine (SVM), kernel machine, and so on. Micro-Doppler signatures can be used as features for classification. Stove and Sykes reported an operational radar system that uses Doppler spectrum for target classification and found that humans, vehicles, helicopters, and ships were successfully classified with multiple Fisher linear discriminators on their Doppler spectra [18, 19]. The success of the Doppler spectrum-based target classification indicates the possibility of using micro-Doppler signatures for target classification.

Micro-Doppler signature-based methods for target classification have been investigated [17, 20–26]. Smith discussed micro-Doppler signature-based classification using an experimental multistatic system [17]. Anderson described the classification using the support vector machine (SVM) and the Gaussian mixture model (GMM) classifiers from micro-Doppler data [20]. Bilik et al. also reported a GMM-based classifier using features of spectral periodicity extracted by the cepstrum coefficients of micro-Doppler data [21]. For classifying human activities, because different movements have various micro-Doppler signatures, by exploiting these differences, a human activity classifier can be formulated. Features used in the human activity classifier may include the torso signature curve, the maximal Doppler shift of the signature, the offset of the signature, the maximum Doppler variation of the torso curve, the oscillation frequency or period of the human locomotion, the kinematic parameters of limbs, and other available features.

However, despite the successful tracking and interpretation of human movements reported in computer vision literature, the selection of suitable descriptive features for prediction of human actions and intention is still a challenge. Most of micro-Doppler signature-based classifiers did not utilize decomposed signatures associated with human body parts and limbs. According to biological motion perception study, if a kinematic information-based classifier is used based on the features of human body parts and limbs extracted from micro-Doppler signatures, this will lead to a more impressive performance for identifying human actions, emotions, and even gender information.

9.3.3   Deep Learning for Micro-Doppler Signature-Based Classification, Recognition, and Identification

Humans and animals can learn to see, perceive, act, and communicate with an efficiency that no machine learning method can approach. The brains of humans and animals are deep in the sense that each action is the result of a long chain of synaptic communications with many layers of processing. Current research on efficient deep learning architectures and on unsupervised learning algorithms can be used to produce deep hierarchies of features for machine learning.

Machine learning is a subset of artificial intelligence (AI) in that a machine demonstrates a sort of intelligence. However, machine learning is a more specific topic in that it gives computers the ability to learn without being explicitly programmed [27].

Deep learning is a subset of machine learning that uses artificial neural networks to learning tasks. It is essentially the attempt at having an artificial brain and nervous system in a machine [28].

Convolutional neural networks (CNNs) are deep, feed-forward artificial neural networks that have successfully been applied to analyzing visual imagery [29]. Based on micro-Doppler signatures, deep learning CNNs have been successfully applied to classifying human activities and fine-grained hand gestures [11, 30].

9.3.4   Aural Methods for Micro-Doppler-Based Discrimination

A challenge in using the micro-Doppler effect in radar is how to effectively communicate the data to an operator. An audio sound of a micro-Doppler embedded signal, called an aural signal, may help a human listener to distinguish between different movements of a target of interest (human walking, running, or jumping) or distinguish between different targets (a wheeled vehicle and a tracked vehicle).

The function of the human auditory classification is based on speech phonemics. A phoneme is a specific sound pattern that can be recognizable by human brains. It is reasonable to generalize this ability of human auditory system for listening to micro-Doppler signals of different movements and to classify target movements using movement phonemes. The human brain neural structures and learned behaviors used on a daily basis are ready to perform micro-Doppler signal processing aurally.

One advantage of the auditory classification systems is that the human auditory classification process is particularly robust in the presence of noise.

Thus, the human auditory system could be an effective alternative in a signal classification system based on movement phonemes. However, the aural micro-Doppler signal is not a conventional speech signal. The human auditory classification system has been optimized for speech signals, but not optimized for aural micro-Doppler signals.

An aural signal converts the micro-Doppler signal to frequencies that a human can hear, which can be used to recognize different movements such as human walking, running, or jumping and to recognize different subjects such as humans or animals.

The aural signal classification is already used in sonar signal classification, but has not yet found widespread use [31, 32]. The potential application of the aural classification to micro-Doppler signatures is also possible to classify a target's different movements by directly converting the baseband micro-Doppler signal into an audio signal for training listeners. However, aural classifiers may not easily classify human actions, emotions, and intentions. They can only serve as a supplementary classifier.

9.3.5 Through-the-Wall Micro-Doppler Signatures

The ability of radar to detect human beings and their movements offers through-the-wall radar applications, including locating living humans after an earthquake or in explosion scenarios, monitoring human activities behind walls, and many other uses. Like micro-Doppler signatures of target captured in an open, free space, micro-Doppler signatures of targets behind walls can also be used to detect and identify the targets behind walls [33]. The effect of targets behind a wall undergoing micromotions has been studied, and the impact of the wall on the micro-Doppler effect has been formulated [34]. It was found that the micro-Doppler effect in the presence of a wall has a similar form as that in free space. However, the measured aspect angle of a target behind a wall is different to that observed in free space. The measured angle depends on the thickness and dielectric constant of the wall. The change of the instantaneous aspect angle due to the wall will affect the radar imaging of the target. However, the presence of a wall does not change the pattern of the micro-Doppler signature of the target; the wall only changes the absolute value of the micro-Doppler signature, depending on the wall properties. Therefore, radar micro-Doppler signatures can be used to detect the presence of human beings and their movements behind the wall.

Through-wall radars typically operate at frequencies lower than 5 GHz. Radar signals that travel through walls will experience attenuation and dispersion. Wall losses can reduce the signal-to-noise ratio (SNR) in radar returns and thus limit the maximum radar detectable range.

Because of the relatively lower frequency used in through-the-wall radars, micro-Doppler frequency shift can be very low. This makes it difficult to detect an object behind the wall. However, with advanced signal decomposition and processing techniques, radar can still sense human body motion, breathing, and heartbeat for detecting and monitoring human activities in cases of finding living humans after an earthquake or in explosion scenarios.

The impact of the wall on the micro-Doppler effect has been investigated. It was found that the presence of a wall does not change the pattern of the micro-Doppler signature of an object. The wall only changes the absolute value of the micro-Doppler signature depending on the wall properties. Therefore, radar micro-Doppler signatures can be used to detect the presence of human beings and their movements behind wall.

9.3.6 Micro-Doppler Signatures for Detection of Targets in Sea Clutter

The detection and tracking of small vessels in sea clutter are important for both civilian and military applications. A small vessel on the sea has its own motions influenced by the sea clutter. The sea clutter is a complex phenomenon influenced by the sea surface and environment such as the sea state, sea waves, sea swell, wind direction, and speed [35–37].

An effective technique that separates a small vessel signature from strong sea clutter is important [38–41]. Therefore, how to utilize the micro-Doppler signatures of small vessels to detect and separate them from sea clutter becomes an interesting research topic [38].

To develop suitable algorithms for detecting and tracking small vessels in sea clutter, it is necessary to characterize sea clutter returns. Sea wave dynamics are determined by the sea state and the direction of sea waves. An example of micro-Doppler signatures of a small fast boat, flying birds, and sea clutter is shown in Figure 9.2. The micro-Doppler signature of the sea clutter appears asymmetric in shape, with a nonzero mean Doppler shift. These features highlight the complexity of the relationship between the intensity modulation and the form of the Doppler frequency distribution. The intensity modulation is dominated by the swell structure in the sea surface and the Doppler frequency distribution is affected by the local gusting of the wind and the detailed scattering mechanism.

The information contents of the small boat in the radar received signals are all embedded in the range profiles. Based on the different motion kinematic

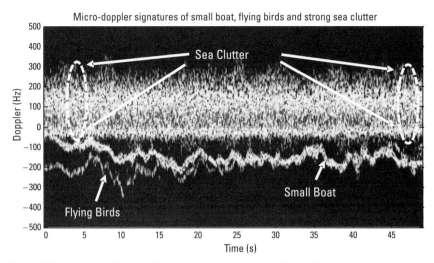

Figure 9.2   Separated micro-Doppler signatures of small boat, flying birds, and strong sea clutter [38].

parameters and different range distributions, the range profiles can be used to extract the micro-Doppler signature of the small boat and other moving objects.

References

[1]   Chen, V. C., D. Tahmoush, and W. J. Miceli, (eds.), *Radar Micro-Doppler Signature: Processing and Applications*, London, U.K.: IET, 2014.

[2]   Zhang, Q., Y. Luo, and Y. A. Chen, *Micro-Doppler Characteristics of Radar Targets*, New York: Elsevier, 2017.

[3]   Amin, M. G., (ed.), *Radar for Indoor Monitoring Detection, Classification, and Assessment*, Boca Raton, FL: CRC Press 2018.

[4]   Chen, V. C., "Advances in Applications of Radar Micro-Doppler Signatures," *2014 IEEE Conference on Antenna Measurements & Applications (CAMA)*, 2014, pp. 1–4.

[5]   Nanzer, J. A., "Millimeter-Wave Interferometric Angular Velocity Detection," *IEEE Transactions on Microwave Theory & Techniques*, Vol. 58, No. 12, 2010, pp. 4128–4136.

[6]   Chatzigeorgiadis, F., and D. Jenn, "A MATLAB Physical-Optics RCS Prediction Code," *IEEE Antenna and Propagation Magazine*, Vol. 46, No. 4, August 2004, pp. 137–139.

[7]   Chatzigeorgiadis, F., "Development of Code for Physical Optics Radar Cross Section Prediction and Analysis Application," Master's Thesis, Naval Postgraduate School, Monterey, CA, September 2004.

[8]   Li, G., et al., "Sparsity-Based Dynamic Hand Gesture Recognition Using Micro-Doppler Signatures," *Proceedings of IEEE 2017 Radar Conference*, 2017.

[9]   Kim, Y. W., and B. Toomajian, "Hand Gesture Recognition Using Micro-Doppler Signatures with Convolutional Neural Network," *IEEE Access*, Vol. 4, 2016, pp. 7125–7130.

[10]   Molchanov, P. et al., "Short-Range FMCW Monopulse Radar for Hand-Gesture Sensing," *Proceedings of IEEE Radar Conference*, 2015, pp. 1491–1496.

[11]   Lien, J., et al., "Soli: Ubiquitous Gesture Sensing with Millimeter Wave Radar," *ACM Transactions on Graphics*, Vol. 35, No. 4, 2016, pp. 142:1–142-19.

[12]   Mallat, S., and Z. Zhang, "Matching Pursuit with Time-Frequency Dictionaries," *IEEE Transactions on Signal Processing*, Vol. 40, No. 12, 1993, pp. 3397–3415.

[13]   Qian, S., and D. Chen, "Signal Representation Using Adaptive Normalized Gaussian Functions," *Signal Processing*, Vol. 36, No. 1, 1994, pp. 1–11.

[14]   Li, P., D. C. Wang, and L. Wang, "Separation of Micro-Doppler Signals Based on Time Frequency Filter and Viterbi Algorithm," *Signal, Image and Video Processing*, Vol. 7, No. 3, 2013, pp. 593–605.

[15]   Mazurek, P., "Estimation of Micro-Doppler Signals Using Viterbi Track-Before-Detect Algorithm," *2017 22nd International Conference on Methods and Models in Automation and Robotics*, 2017, pp. 898–902.

[16]   Chernyak, V. S., *Fundamentals of Multisite Radar Systems*, London, U.K.: Gordon and Breach Scientific Publishers, 1998.

[17]   Smith, G. E., "Radar Target Micro-Doppler Signature Classification," PhD Dissertation, Department of Electronic and Electrical Engineering, University College London, 2008.

[18]   Stove, A. G., and S. R. Sykes, "A Doppler-Based Automatic Target Classifier for a Battlefield Surveillance Radar," *2002 International Radar Conference*, Edinburgh, U.K., 2002, pp. 419–423.

[19]   Stove, A. G., and S. R. Sykes, "A Doppler-Based Target Classifier Using Linear Discriminants and Principal Components," *Proceedings of the 2003 International Radar Conference*, Adelaide, Australia, September 2003, pp. 171–176.

[20]   Anderson, M. G., "Design of Multiple Frequency Continuous Wave Radar Hardware and Micro-Doppler Based Detection and Classification Algorithms," Ph.D. Dissertation, University of Texas at Austin, 2008.

[21]   Bilik, I., J. Tabrikian, and A. Cohen, "GMM-Based Target Classification for Ground Surveillance Doppler Radar," *IEEE Transactions on Aerospace and Electronic Systems*, Vol. 42, No. 1, 2006, pp. 267–278.

[22] Zabalza, J., and C. Clemente, "Robust PCA Micro-Doppler Classification Using SVM on Embedded Systems," *IEEE Transactions on Aerospace and Electronic Systems*, Vol. 50, No. 3, 2014, pp. 2304–2310.

[23] Vishwakama, S., and S. S. Ram, "Dictionary Learning for Classification of Indoor Micro-Doppler Signatures Across Multiple Carriers," *2017 IEEE Radar Conference*, 2017, pp. 0992–0997.

[24] Vishwakama, S., and S.S. Ram, "Classification of Multiple Targets Based on Disaggregation of Micro-Doppler Signatures," *2016 Asia-Pacific Microwave Conference*, 2016, pp. 1–4.

[25] Bjorklund, S., H. Petersson, and G. Hendeby, "Features for Micro-Doppler Based Activity Classification," *IET Radar, Sonar and Navigation*, Vol. 9, No. 9, 2015, pp. 1181–1187.

[26] De Wit, J. J. M., R. Harmanny, and P. Molchanov, "Radar Micro-Doppler Feature Extraction Using the Singular Value Decomposition," *Proceedings 2014 International Radar Conference*, Lille, France, 2014, pp. 1–6.

[27] Samuel, A., "Some Studies in Machine Learning Using the Game of Checkers," *IBM Journal of Research and Development*, Vol. 3, No. 3, 1959, pp. 210–229.

[28] McCulloch, W. S., and W. Pitts, "A Logical Calculus of the Ideas Immanent in Nervous Activity," *Bull. Math. Biophys.*, Vol. 5, 1943, pp. 119–133.

[29] LeCun, Y., and Y. Bengio, "Convolutional Networks for Images, Speech, and Time-Series," in M. A. Arbib, editor, *The Handbook of Brain Theory and Neural Networks*. Cambridge, MA: MIT Press, 1995.

[30] Kim, Y., and T. Moon, "Human Detection and Activity Classification Based on Micro-Doppler Signatures Using Deep Convolutional Neural Networks," *IEEE Geoscience and Remote Sensing Letters*, Vol. 13, No. 1, 2016, pp. 8–12.

[31] Hines, P. C., and C. M. Ward, "Classification of Marine Mammal Vocalizations Using an Automatic Aural Classifier," *J. Acoust. Soc. Am.*, Vol. 127, No. 1970, 2010.

[32] Allen, N., et al., "Study on the Human Ability to Aurally Discriminate Between Target Echoes and Environmental Clutter in Recordings of Incoherent Broadband Sonar," *J. Acoust. Soc. Am.*, Vol. 119, No. 3395, 2006.

[33] Chen, V. C., et al., "Radar Micro-Doppler Signatures for Characterization of Human Motion," Chapter 15 in *Through-the-Wall Radar Imaging*, M. Amin, (ed.), Boca Raton, FL: CRC Press, 2010.

[34] Liu, X., H. Leung, and G. A. Lampropoulos, "Effects of Non-Uniform Motion in Through-the-Wall SAR Imaging," *IEEE Transactions on Antenna and Propagation*, Vol. 57, No. 11, 2009, pp. 3539–3548.

[35] Watts, S., "Modeling and Simulation of Coherent Sea Clutter," *IEEE Transactions on Aerospace and Electronic Systems*, Vol. 48, No. 4, 2012, pp. 3303–3317.

[36]   Spyrosm, P., and J. S. John, "Small-Target Detection in Sea Clutter," *IEEE Transactions on Geoscience and Remote Sensing*, Vol. 42, No. 7, 2004, pp. 1355–1361.

[37]   Javier, C. M., "Statistical Analysis of a High-Resolution Sea-Clutter Database," *IEEE Transactions on Geoscience and Remote Sensing*, Vol. 48, No. 4, 2010, pp. 2024–2037.

[38]   Chen, V. C., and D. Tahmoush, "Micro-Doppler Signatures of Small Boats in Sea," in Chapter 10, *Radar Micro-Doppler Signature: Processing and Applications*, V. C. Chen, D. Tahmoush, and W. J. Miceli, (eds.), Radar Series 34, IET, 2014, pp. 345–381.

[39]   Chen, X. L., J. Guan, and Y. He., "Detection and Estimation Method for Marine Target with Micromotion Via Phase Differentiation and Radon-LV's Distribution," *IET Radar, Sonar & Navigation*, Vol. 9, No. 9, 2015, pp. 1284–1295.

[40]   Chen, X. L., et al., "Detection and Extraction of Target with Micromotion in Spiky Sea Clutter Via Short-Time Fractional Fourier Transform," *IEEE Transactions on Geoscience and Remote Sensing*, Vol. 52, No. 3, 2014, pp. 1002–1018.

[41]   Chen, X. L., et al., "Detection of Low Observable Sea-Surface Target with Micromotion Via Radon-Linear Canonical Transform," *IEEE Geosci. Remote Sens. Letter*, Vol. 11, No. 7, 2014, pp. 1225–1229.

About the Author

Victor C. Chen, a Life Fellow of the Institute of Electrical and Electronics Engineers (IEEE), is internationally recognized for his work on the micro-Doppler effect in radar and time-frequency-based radar image formation. He received his Ph.D. in electrical engineering from Case Western Reserve University, Cleveland, Ohio. In 1990, he joined the Radar Division at the U.S. Naval Research Laboratory. After retiring from the U.S. Naval Research Laboratory in 2010, he cofounded and serves as the chief technologist at Ancortek, Inc., in the United States. He authored numerous papers in journals, proceedings, and books, including four books: *Time-Frequency Transforms for Radar Imaging and Signal Analysis* in 2002, *The Micro-Doppler Effect in Radar* in 2011, *Radar Micro-Doppler Signatures: Processing and Applications* in 2014, and *Inverse Synthetic Aperture Radar Imaging: Principles, Algorithms and Applications* in 2014.

Index

Time-Frequency Signal Analysis with Applications, Ljubiša Stanković, Miloš Daković, and Thayananthan Thayaparan

Time-Frequency Transforms for Radar Imaging and Signal Analysis, Victor C. Chen and Hao Ling

Transmit Receive Modules for Radar and Communication Systems, Rick Sturdivant and Mike Harris

For further information on these and other Artech House titles, including previously considered out-of-print books now available through our In-Print-Forever® (IPF®) program, contact:

Artech House	Artech House
685 Canton Street	16 Sussex Street
Norwood, MA 02062	London SW1V HRW UK
Phone: 781-769-9750	Phone: +44 (0)20 7596-8750
Fax: 781-769-6334	Fax: +44 (0)20 7630-0166
e-mail: artech@artechhouse.com	e-mail: artech-uk@artechhouse.com

Find us on the World Wide Web at: www.artechhouse.com